FRANKLIN & WASHINGTON

ALSO BY EDWARD J. LARSON

To the Edges of the Earth: 1909, the Race for the Three Poles, and the Climax of the Age of Exploration

The Return of George Washington: Uniting the States, 1783–1789

An Empire of Ice: Scott, Shackleton, and the Heroic Age of Antarctic Science

A Magnificent Catastrophe: The Tumultuous Election of 1800, America's First Presidential Campaign

Summer for the Gods: The Scopes Trial and America's Continuing Debate over Science and Religion

Evolution: The Remarkable History of a Scientific Theory

Trial and Error: The American Controversy over Creation and Evolution

Evolution's Workshop: God and Science on the Galapagos Islands

Sex, Race, and Science: Eugenics in the Deep South

George Washington, Nationalist

The Constitutional Convention: A Narrative History from the Notes of James Madison (with Michael Winship)

On Faith and Science (with Michael Ruse)

FRANKLIN & WASHINGTON

The Founding Partnership

EDWARD J. LARSON

WM

WILLIAM MORROW
An Imprint of HarperCollinsPublishers

Grateful acknowledgment is made to the following for use of the images that appear in the art insert: the National Portrait Gallery (page 1, top); the Mount Vernon Ladies' Association (page 1, bottom; page 2, top and bottom; page 4, top; page 10, bottom; page 11, bottom); the New York Public Library (page 3, top; page 6, bottom left); Flickr (page 3, bottom); the Library of Congress (page 5, bottom left; page 8, top; page 9, bottom; page 11, top; page 13, top; page 14, top); and the Architect of the Capitol (page 7; page 12, bottom; page 16).

HarperCollins books may be purchased for educational, business, or sales promotional use. For information, please e-mail the Special Markets Department at SPsales@harpercollins.com.

FIRST EDITION

Designed by Bonni Leon-Berman

Library of Congress Cataloging-in-Publication Data has been applied for.

ISBN 978-0-06-288015-4

20 21 22 23 24   LSC   10 9 8 7 6 5 4 3 2 1

CONTENTS

PREFACE

"My Dear Friend"

ON MAY 13, 1787, one day before the scheduled start of the Constitutional Convention, Virginia delegate George Washington arrived in Philadelphia and, as his first formal act, visited Benjamin Franklin at his home. Everyone knew Washington would chair the Convention and, if it succeeded, lead the nation. Franklin then served as president of Pennsylvania, the host state, and stood as the only American with stature comparable to Washington's. By all accounts, they were the two indispensable authors of American independence and key partners in any attempt to craft a more perfect union at the Convention.

Although they had different views on what was needed in a new constitution, both men agreed that fundamental reform was essential. Less than a month earlier, Franklin wrote to Thomas Jefferson about the Convention, "If it does not do Good it must do Harm, as it will show that we have not Wisdom enough among us to govern ourselves."[1] And after the Convention, Washington repeatedly warned that without the Constitution, political chaos would engulf the land.[2] Franklin and Washington had staked their lives and fortunes on the American experiment in liberty and were committed to its preservation.

Escorted into Philadelphia by a military guard and greeted with the ringing of bells and cheers of townspeople, Washington no sooner unloaded his luggage at Robert Morris's mansion, where he was staying, than he called upon Franklin. Some historians surmise that, for added dignity, Washington rode the two blocks between the Morris and Franklin homes on Market Street in his fashionable carriage. I suspect he walked. The distance was almost too short to ride, Washington enjoyed walking, and any visit from the general carried dignity. More critically, riding would mean turning into the newly constructed street-front arch and arriving at Franklin's interior courtyard with an enslaved Black coachman and two liveried slaves serving as footmen. Seven years earlier, Quaker Pennsylvania became the first state to end slavery by statute and, by the time of Washington's visit, Franklin presided over the state's leading abolition society. Among his many gifts, Washington was an astute politician and pitch-perfect political actor who knew how to forge alliances. Arriving by carriage would not impress the down-to-earth Franklin and the attendance of slaves might offend him. Having known Franklin for more than thirty years, Washington would have understood how to greet him—just as Franklin would know how to welcome Washington.

Although not as accomplished an inventor or scientist as Franklin—no one in America was—Washington had an enlightened mind that embraced inventions and appreciated science. While it was modest in comparison to Washington's Mount Vernon estate, Franklin's recently remodeled house included features certain to delight Washington. An enlarged second-floor library held more than four thousand books, making it one of the largest private collections in America; a glass armonica of Franklin's own design, for which he composed music; and his already fabled electrical equipment. "I hardly know how to justify building a Library at an Age that will so soon oblige me to quit it," Franklin, then eighty, had written to his sister six months earlier but by now had probably forgotten such reservations.[3] Boyish in his enthusiasm despite physical infirmities, Franklin also likely showed Washington a mechanical arm used to retrieve books from upper shelves, an early type of copying machine that employed slow-drying ink and a press to duplicate newly written letters, and a reading chair with foot pedals to propel a fan. Few visitors would have appreciated such curiosities more than the mechanically minded Washington. Indeed, he purchased a fan chair while in Philadelphia for use at his Mount Vernon library.

After the house tour, Franklin likely discussed prospects for the Convention with Washington over tea or wine. It would have been tea in Franklin's youth, but he had developed a taste for fine wines while serving in Paris during the Revolution. Washington also enjoyed wine. They would reunite for dinner at Franklin's house three days later, when Franklin opened a cask of dark English beer to everyone's delight, but wine was more fitting for an afternoon visit. Perhaps reserving the large, remodeled dining room for the later dinner, this talk probably occurred at an outdoor table set under a large mulberry tree in Franklin's garden. Following his return from Europe, Franklin had turned this space from growing vegetables into a flower garden that would have been in full bloom by mid-May. Over the summer, he frequently entertained Convention delegates there. This would have been its first such use, and perhaps the most important. Franklin and Washington would go out from there to forge a nation from thirteen states.

"MY DEAR FRIEND" were the last words that Benjamin Franklin addressed to George Washington. They came at the end of a letter written in what Franklin knew would be his final year of life. Washington closed his response to Franklin with the salutation "Your sincere friend." In this exchange, written in the first year of Washington's presidency, each expressed his undying "respect and affection" for the other, with Franklin adding "esteem" and Washington topping him with "veneration." At the time, Franklin and Washington were the two most admired individuals in the United States, and the most famous Americans in the world.

Their final letters to each other represented a fitting end to a three-decade-long partnership that, more than any other pairing, would forge the American nation. Their relationship began during the French and Indian War, when Franklin supplied the wagons for British general Edward Braddock's ill-fated assault on Fort Duquesne and Washington buried the general's body under the dirt road traveled by those retreating wagons. Both had warned Braddock against the frontier attack. Rekindled in 1775 during the Second Continental Congress, this friendship continued through the American Revolution, the Constitutional Convention, and the establishment of the new federal government. Perhaps because of differences in their background, age, manner, and public image, their relationship was not widely commented on then, and it remains little discussed today. But it existed, and helped to shape the course of American history. Both men have been called "the first American," but they were friends first and, unlike John Adams and Thomas Jefferson, never rivals.

Their relationship gained historical significance during the American Revolution, when Franklin led America's diplomatic mission in Europe and Washington commanded the Continental Army. Victory required both of these efforts to succeed, and their success required coordination and cooperation. No less an authority than Jefferson testified to the success of their efforts when, after the Revolution, he observed that, in terms of their contributions to the patriot cause, the world had drawn a broad line between Washington and Franklin "on the one side, and the residue of mankind on the other."[4] Their successful collaboration during the Revolution, especially when coupled with their role as two of the most prominent delegates at the Constitutional Convention, helped to found a nation and propel a global experiment with individual liberty and republican rule.

Leadership at this level is a rare quality and well worth study. Leadership studies, however, typically focus on individuals, either singly or in comparative analysis. Pairs or teams are less often the center of study unless they collaborate in some formal institutionalized manner, such as members of a cabinet or business partners. Franklin and Washington never had official ties, yet they worked together toward a common cause with extraordinary success. To explore their historic collaboration, this book traces their shared history in a dual biography that looks for overlaps and stresses connections.

Despite differences in public image and private style, striking similarities emerge. Both men had successful business and political careers in late colonial America and led their respective colonies' defense during the French and Indian War. After long supporting royal rule in the colonies, both became key early proponents of independence. Both then sought to strengthen the union of the states, leading to the framing and ratification of the Constitution. Instinctively, each worked to forge consensus and lead through others.

In the United States, leadership studies often privilege the founders, with the bias increasing in times, such as now, when founding institutions appear under stress. What would Washington, or Franklin, or Alexander Hamilton, or James Madison say or do in these times? Of course, any such quest is ahistorical, and not the object of this book. Yet current events inevitably color history as it is written and read, no matter how much the writer or the reader seeks to avoid or minimize it. The endless flow of books about either Franklin or Washington—and no American save Abraham Lincoln has generated more—typically do not deal extensively with their relationship. Biographers are much more likely to pair Washington and Hamilton or Jefferson and Madison or Franklin and Jefferson than Franklin and Washington. As I researched and wrote this book, however, I was repeatedly surprised by the extent of the links between them. As I proceeded, each man became more understandable in light of the other.

In a sense, it became clear that I had been working on this book throughout my academic career. At Williams College, presidential historian James MacGregor Burns introduced me to leadership studies, and, as his research expanded to include Washington, I benefited from his collaboration with his partner and my friend Susan Dunn. In graduate school at the University of Wisconsin, I studied the founders with Norman Risjord, Paul Boyer, and (for Franklin's science) Ronald Numbers. Law school brought study of the Constitution with Laurence Tribe, John Hart Ely, and others, with opportunities to cross Harvard Yard and learn about Franklin with historian of science I. Bernard Cohen. Various teaching opportunities and research fellowships deepened my appreciation of Franklin, Washington, and the founders, including terms at the Fred W. Smith National Library for the Study of George Washington at Mount Vernon, holding the John Adams Chair in American Studies in the Netherlands and the Douglas Southall Freeman Chair in history at the University of Richmond, and teaching at Stanford University with legal historian Lawrence Friedman and in proximity to Madison scholar Jack Rakove and historian of the American Revolution Ray Raphael.

I am particularly thankful that for this book, I had the inestimable advantage of working with the same editor at HarperCollins, Peter Hubbard, as for my earlier monograph on Washington. This time, our manuscript benefited from meticulous copy editing by

Trent Duffy, who brought with him local knowledge as a native of Morristown, New Jersey. I also wish to thank archivists and research librarians at UCLA's Charles E. Young Research Library, the Library of Congress, the National Archives, the Pepperdine University libraries, Stanford's Cecil H. Green Library, the American Antiquarian Society, and the American Philosophical Society and Mary Thompson at Mount Vernon. When I began studying Franklin and Washington, the incomplete status of the extraordinarily ambitious projects to collect and publish their papers frustrated me. Over the years, however, I grew to anticipate and then enjoy each volume in those series as they appeared. I am still not sure if I will live to see the end of these projects as they grind through successive years of letters, papers, and publications—but if I do, I will miss the anticipation and discovery that comes with each succeeding volume. I would like to dedicate this book to those projects. They, more than any other single resource, made this work possible.

I close this preface, and with it the writing of this book on founding friends, as I look over Lake Geneva not far from Château d'Hauteville, the country seat of Pierre-Philippe Cannac, who installed Franklin's lightning rods on the château when he built it during the 1760s. In 1794, the château passed through marriage and inheritance to Daniel Grand, the youngest son of Franklin's French banker Rodolphe-Ferdinand Grand, whose descendants live there to this day. Much of Franklin's correspondence with the father moved to the château with the son. The now 250-year-old lightning rods—the first installed in Switzerland—still protect the château from the thunderstorms that rake the region, much as Franklin's works and example during the founding era, along with those of Washington, help the American people weather the political storms that at times beset them. From here, when the weather clears, I can see the sun set over the lake better than I can see it rise, but, like Franklin at the Constitutional Convention, I hope that it is still rising over the American experiment in liberty.

EJL
Lausanne, Switzerland

Book I

CONVERGING LIVES

One

GREAT EXPECTATIONS

THEIRS ARE THE TWO most recognizable portraits ever painted of any Americans. Benjamin Franklin, his body turned left but face looking forward, appears eminently approachable, with a slight grin; loose, flowing hair; and a twinkle in his large, soft eyes. He might have just cracked a joke or told a witty remark to the painter, Joseph Duplessis. George Washington, his shoulders turned right in the original but also facing forward, looks coldly statuesque—a marble bust—with tight, drawn lips; formal, powdered hair; and narrow, piercing eyes. He had rebuked the painter, Gilbert Stuart, for suggesting a more informal pose.

"Now sir, you must let me forget that you are General Washington," the renowned portraitist breezily offered.

You need not forget "who General Washington is," came the sitter's terse reply.[1]

These images adorn America's two most widely circulated banknotes: Stuart's 1796 Athenaeum portrait of Washington (flipped to turn left) on the ubiquitous American one-dollar bill and Duplessis's 1785 painting of Franklin on the widely hoarded hundred. Together, they account for more than three-fifths of all U.S. banknotes in circulation, with "Franklins" constituting some 80 percent of the total value of all American paper money.

Franklin's portrait expands beyond the bill's borders—his expressive face looking fleshy and open. Clearly aged, Franklin nevertheless appears vibrant and alive. "He . . . possesses an activity of mind equal to a youth of 25," a fellow delegate said of the eighty-one-year-old Franklin at the Constitutional Convention, two years after the picture was painted.[2] Washington, in contrast, glares out from a tight central oval on the one-dollar bill, his craggy, colorless visage looking like a plaster mask animated only by those intense eyes, which Stuart painted as bluer than they really were. Once told that his expression showed emotion, Washington shot back, "You are wrong. My countenance never yet betrayed my feelings!"[3] Stuart saw something in that face, however. "All his features," the painter commented, "were indicative of the strongest and most ungovernable passions."[4] Yet Washington governed them with a granite, tight-lipped self-control that made him the stoic father figure for a nation that adopted Franklin as its favorite uncle. Together, they midwifed a republic.

Stamped on the national consciousness as the definitive representation of the episode it captures, Howard Chandler Christy's *Scene at the Signing of the Constitution of the United States* conveys the enduring sense that Franklin and Washington stood at the crossroads of American history and shaped its course. Commissioned by Congress during the Great Depression, this historical painting artfully arranges the Constitution's thirty-seven other signers in imaged poses around Franklin and Washington inside Philadelphia's Independence Hall. Of all the delegates, only Franklin, seated at center, looks directly forward, as if to engage modern viewers. Alexander Hamilton leans in to catch his ear. Half obscured at Franklin's side, James Madison looks up toward Washington, who stands in near profile on a dais to the viewer's right, head and shoulders above the other signers as he gazes beyond them, as if to some distant shore.

Here too, as in their individual portraits, the blue-garbed Franklin appears warm and gregarious while the black-suited Washington looks cold and aloof. Ignoring their placement on the canvas, the painting's official key, supplied to identify those pictured, designates Washington by the number 1, Franklin by 2, Madison 3, and Hamilton 4. Others follow in an order that seemingly reflects their relative contributions to the nation's founding. If so, then pairing Franklin and Washington at top perpetuates the popular view.

Of all the public art linking and extolling the two principal founders, surely the most dramatic is *The Apotheosis of Washington*. Situated in the concave apex or "eye" of the U.S. Capitol's rotunda, the painting depicts Washington in flowing robes rising to heaven flanked by the goddesses of liberty and victory along with thirteen classically garbed maidens. A rainbow descending toward earth from Washington's feet alights on the shoulder of Franklin, who is toiling in the garden of science and invention. At more than twice their actual heights and looking fit, Washington and Franklin are the largest historical figures in the picture and portrayed so as to be recognizable by viewers looking upward from some 180 feet below on the rotunda floor. Painted in the true fresco technique by Vatican artist Constantino Brumidi in 1865 at the end of the Civil War, the circular canopy, which is more than 200 feet in circumference, captures the historical consensus of a remote godlike Washington and a pragmatic earth-bound Franklin towering over the nation's founding.

Not all historians speak kindly about Franklin and Washington, but few modern scholars question their preeminence among the founders. "Of those patriots who made independence possible, none mattered more than Franklin, and only Washington mattered as much," H. W. Brands observed in 2000.[5] That same year, Joseph Ellis, in his acclaimed book on the revolutionary generation *Founding Brothers,* put Franklin and Washington atop his short list of leaders who stepped forward at the national level to promote the cause of independence when doing so was still perilous and hailed them as the new republic's two most influential political figures.[6] Garry Wills expressed much the same view in 2002, as did Gordon Wood in 2004.[7] Writing of Franklin, Wood concluded, "His critical diplomacy in the Revolution makes him second only to Washington."[8]

Such assessments are not new. "The History of our Revolution will be one continued Lye from one End to the other," Vice President John Adams famously complained in 1790. "The Essence of the Whole will be *that Dr Franklins electrical Rod, Smote the Earth and out Spring General Washington. That Franklin electrified him with his Rod—and thence forward these two conducted all the Policy Negotiations Legislation and War.* These underscored Lines contain the whole Fable Plot and Catastrophy."[9]

Perhaps because they are remembered so differently, however, except in Adams's rant and Christy's painting, the two are not generally portrayed together. This clouds our understanding of both men. They worked shoulder to shoulder on the patriot cause, and not only at the Constitutional Convention in 1787. Indeed, they came to the Convention as longtime friends.

RESIDING IN TWO COLONIES that then asserted overlapping claims to the western frontier, the remarkable relationship between Pennsylvania's Franklin and Virginia's Washington began three decades before the Convention, when each led their respective colonies' militia during the French and Indian War. By then Franklin was fifty, and one of the most widely known and respected persons in the Western world. Washington was then less than half Franklin's age but already held a regional reputation for military prowess. Neither man was born to power or influence; both had earned it.

In 1756, Pennsylvania and Virginia were two of England's most prosperous and populous American colonies. They had overlapping claims to western lands on the Ohio frontier, where the French and Indian War began and was initially fought. Those early battles went badly for the English side. That these two colonies turned to Franklin and Washington in desperate times speaks to the standing that they had gained long before the American Revolution made them into national heroes and global icons of democracy and freedom.

During the 1750s, Franklin's international reputation rested on his scientific achievements in electricity, but he was best known locally as a printer, writer, civic reformer, postmaster, and pragmatic political leader. Given his humble origins, this represented a stunning achievement for the time and remains a lasting testament to his genius.

Born the fifteenth of seventeen children (ten of whom were alive at the time) to a working-class family in Puritan Boston on January 17, 1706, Franklin was the youngest son of his father's second wife. Bookish and inquisitive, he was apprenticed at age twelve to his older brother James, a local printer whose products soon included an independent weekly newspaper, the *New-England Courant*. The younger Franklin quickly displayed a seemingly inexhaustible capability for hard work and, in large part self-taught by reading the essays of Joseph Addison and Richard Steele, an acquired facility to write wry, instructional, and satirical prose. Without telling his brother, Franklin began contributing letters to the newspaper under the female pseudonym Silence Dogood. "I am an Enemy to Vice, and a Friend to Vertue," he had her observe. "I am courteous and affable, good humour'd (unless I am first provok'd,) and handsome, and sometimes witty." Being "a mortal Enemy to arbitrary Government and unlimited Power," she added, "I am naturally very jealous for the Rights and Liberties of my Country; and . . . never intend to wrap my Talent in a

Napkin."[10] This was Franklin. Only sixteen, he knew himself well and could express it from another's viewpoint.

Having mastered the printer's trade by age seventeen, in 1723 Franklin fled what he saw as an unduly oppressive apprenticeship to his brother and the stifling religiosity of Boston for New York. Finding no work there, he went on to the burgeoning Quaker community of Philadelphia, a place of relative religious tolerance, political freedom, and economic opportunity, where he heard that a printer's assistant was needed. Arriving with little money and nothing to recommend him for employment beyond his talent for setting type, Franklin found part-time work with the town's newly established second printer, Samuel Keimer.

Neither of Philadelphia's two printers was qualified for their trade, Franklin soon concluded, while he impressed others with his skill and work ethic—so much so that the colony's governor offered to set him up in his own shop. Only after arriving in London to select the needed equipment did Franklin learn that the governor's credit was no good. He worked for eighteen months in various English print shops before returning to Philadelphia with a Quaker merchant who offered him work as a salesclerk. When the merchant died after only a few months, Franklin found himself back working for Keimer before teaming with a coworker to open their own print shop and then buying out his partner to have a shop of his own. Our "industry visible to our Neighbors began to give us Character and Credit," Franklin said of the partnership, and it proved even truer when he worked alone.[11] "I took care not only to be in *Reality* Industrious and frugal, but to avoid all *Appearances* to the contrary," he wrote of his early days as a sole proprietor, "and, to show that I was not above my Business, I sometimes brought home the Paper I purchas'd at the Stores, thro' the Streets on a Wheelbarrow."[12] This proved the way to wealth in Quaker

Philadelphia. Soon Franklin dominated the regional printing trade and began branching out to related businesses.

Nearly six feet tall with an athletic build that became paunchy with age, Franklin was a strong swimmer and able equestrian who believed in the benefits of vigorous exercise and fresh air. On his return voyage from England, for example, he worked out by swimming around the ship as it sailed at sea, and, at age seventy, he infuriated a traveling companion by insisting on sleeping with the window open. On both health and moral grounds, Franklin experimented with a vegetarian diet during his youth and re-peatedly returned to one in later years for his health, but gave up strict adherence after seeing small fish in the gut of a large, freshly caught cod being filleted for dinner. "If you eat one another, I don't see why I mayn't eat you," he reasoned. "So convenient a thing it is to be a *reasonable Creature,* since it enables one to find or make a Reason for everything one has a mind to do."[13]

While not doubting the existence of God but convinced that acts mattered more than beliefs, Franklin rejected the Christian doctrine of salvation by grace and with it faith in the divinity of Christ. "Morality or Virtue is the End, Faith only a Means to ob-tain that End: And if the End be obtained, it is no matter by what Means," he wrote in 1735.[14] Fundamentally pragmatic in thought, word, and deed, Franklin soon added, "Sin is not hurtful because it is forbidden but it is forbidden because it's hurtful."[15] Regarding the Bible, he concluded, "Tho' certain Actions might not be bad because they were forbidden by it, or good *because* it commanded them; yet probably those Actions might be forbidden *because* they were bad for us, or commanded *because* they were beneficial to us, in their own Natures, all the Circumstances of thing considered."[16] Although not necessarily their creed, apparently this was good enough for the people of Philadelphia. Among them he prospered.

For the next two decades, to 1747, Franklin was consumed by business and local civic affairs. As a printer, he published a newspaper, the *Pennsylvania Gazette,* and an annual almanac, *Poor Richard's.* Filled with witty commentary and practical advice, the *Gazette* became the colony's leading paper and *Poor Richard's* (written under the pseudonym Richard Saunders) gained a wide readership. "The Way to Wealth, if you desire it, . . . depends chiefly on two Words, INDUSTRY and FRUGALITY; i.e. Waste neither Time nor Money, but make the best Use of both," Franklin advised fellow tradesmen.[17] He practiced what he preached. Integrating forward and backward from the printing business, he owned or had an interest in some two dozen print shops in other colonies and almost as many paper mills, championed the issuance of paper money that he subsequently was paid to print, won the contract to print legislative documents for Pennsylvania, became its postmaster to facilitate delivery of his newspaper, and served as clerk of its assembly to gain a leg up in getting the news. By the 1740s, Franklin was serving as comptroller of the post office for British North America and in the next decade became its joint deputy postmaster general.

In 1730, Franklin married the frugal and hardworking Deborah Read, who promptly took in and reared his son, William, by a premarital affair. They had two children together, Francis and Sarah, but the first died at age four from smallpox. "It is the Man and Woman united that make the compleat human Being," Franklin later noted. "If you get a prudent healthy Wife, your Industry in your Profession, with her good Economy, will be a Fortune sufficient."[18] He would trust his own way to wealth without marrying into money and (with his own industry and his wife's economy) became one of the richest American colonists living north of the Mason-Dixon Line.[19] Yet wealth was never his chief goal in life. "The Years roll round, and the last will come," Franklin wrote to his mother, "when I would rather have it said, *He lived usefully,* than, *He died rich.*"[20]

Accordingly, even as his businesses expanded, Franklin threw himself into local civic affairs. An inveterate joiner with a gift for drawing in others, he founded a self-improvement club for up-and-coming tradesmen called the Junto, a secular subscription library, a firefighting brigade, an academy that grew into the University of Pennsylvania, an intercolonial philosophical society, and (in 1747, during King George's War against France and its Native American allies) a ten-thousand-man volunteer militia to defend Pennsylvania when its pacifist Quaker leaders would not. Franklin deeply believed in the power of collective action guided by reason. In 1731, he joined the nascent modern Freemason movement, which then served as something of an Enlightenment era alternative to organized religion by supplying fellowship, secret ritual, and a shared moral code for its members. Rising quickly through the ranks, Franklin, much like Washington twenty years later in Virginia, soon became a leader of his colony's Freemasons—a status that linked him for life with fellow Masons in America and Europe.

HAVING ACHIEVED FINANCIAL INDEPENDENCE, Franklin "retired" in 1748 at age forty-two to a life of public service even as his established businesses and investments continued to generate substantial income. Some sixty years earlier, King Charles II had given William Penn a royal charter to establish a propriety colony for Penn's pacifist Quaker sect and others on a broad stretch of land running from the Delaware River west for five degrees of longitude and north to south from New York to Maryland. As Penn peacefully secured the land from its Native inhabitants by purchases and resold it to European settlers at a handsome profit, the colony flourished.

After Penn's death in 1718, however, tensions developed between his descendants, the colony's successor proprietors (who lived in London, appointed the resident governor, retained vast landholdings in the colony on which they paid no taxes, and drifted from the Quaker faith), and the colonists represented by their Quaker-controlled local assembly. Franklin stepped into the midst of this dispute on the side of his fellow colonists when they elected him as a Philadelphia city councilor in 1748 and a member of the colonial assembly three years later. "I conceiv'd my becoming a Member would enlarge my Power of doing Good," he later wrote, and he threw himself into that task by pushing proposals to improve city streets and public safety.[21]

While these activities enhanced his local reputation, Franklin gained international renown through science. Open-minded and curious, he could see fundamental relationships in nature that eluded others. By comparing the times when storms hit various places, for example, Franklin was first to realize that so-called nor'easters, which featured fierce winds from the north, actually moved from south to north. Expanding on this observation, he advanced the idea that weather moved in predicable patterns that could be forecast. Likewise, Franklin deduced from the relative speeds of transatlantic crossings at various latitudes and directions that currents circulated within oceans—another first. By separating the smoke from the hot air generated in woodstoves and using the radiant power of metal heated by circulating the hot air, he designed a fireplace insert that, while not perfect, pointed the way toward future advances in indoor heating. Ultimately, he made his name by revolutionizing the scientific understanding of electricity.

People had experienced static electric sparks from time immemorial but no one systematically investigated them until the seventeenth century. These early experimenters found that they could generate an electric charge by rubbing various objects, such as a glass rod or amber stone, and collect it in a glass bottle lined and coated with metal foil, called a Leyden jar for where it was developed. Two types of electricity seemed to exist, depending on the type of substance rubbed to create it. Parlor games involving electrical shocks and charges gained popularity in Europe during the 1700s and passed to America by 1747, when Franklin's subscription library acquired equipment for performing electrical experiments from its London agent, the wealthy science devotee Peter Collinson.

Armed with this equipment and unencumbered by preconceived notions about the nature of electricity, Franklin conducted a series of brilliantly designed and easily explained experiments involving people on insulating wax blocks passing and receiving charges. From these experiments, he induced that there is only one type of electricity, that rubbing does not create it but rather causes it to flow, and that positive and negative charges result from a relative surplus or deficit of this universal "Electrical fluid."[22] Here was science fitting the British model decreed by Francis Bacon, and a first step toward both the modern physics of energy conservation and the transforming technology of electrical circuits, all flowing from a colonial tradesman with little formal education. The London stage could not have told a more compelling story and, in 1751, when Collinson published in a small book letters from Franklin describing his discoveries, it quickly went through multiple European editions. "He found electricity a curiosity," Franklin biographer Carl Van Doran concluded, "and left it a science."[23]

Following up on his initial insights, Franklin devised an experiment to test the supposition that lightning was a static electric discharge. That test involved drawing a charge down from storm clouds by a pointed metal conductor—he used a sharp wire extended from a kite—and channeling it into a Leyden jar where its properties could be compared with those of a standard static electric charge. They proved the same. Franklin also improved on the

Leyden jar to create the first electric battery and, always looking for useful application of his discoveries, developed the lightning rod to protect structures by drawing off atmospheric electricity and carrying it safely to earth through a grounding wire.

To Franklin, the neutral balance of electricity on earth, and its tendency to find equilibrium, testified to divine design in nature. The ability of people to use their reason to improve on natural processes, such as through a lightning rod, gave humans a distinct role and purpose in making the world better. In his *Autobiography*, when Franklin proposed the ideal prayer, it was for "that Wisdom that discovers my truest Interests" rather than for any particular blessing to miraculously happen.[24] Franklin's God acted through people. And for his discoveries in electricity, the greatest philosopher of the day, Immanuel Kant, hailed Franklin as a "new Prometheus"—a living demigod—who snatched fire from the heavens and gave it to humanity.[25]

When the French and Indian War broke out between English settlers and French and their Native allies on the Pennsylvania and Virginia frontier in 1754, frightened Pennsylvanians inevitably turned to Franklin for leadership. Equally inevitably, he provided it. This threw him together with Washington, his Virginia counterpart.

AS A MEMBER OF VIRGINIA'S LANDED GENTRY, George Washington was born with higher social status than Franklin at a time and place where such things mattered. And yet, all factors considered, his prospects were hardly greater. Although Franklin's father was poorer than Washington's, he was a respected member of his colony's established church, which created opportunities for his sons. Had Franklin been more pious, he would likely have gone to Harvard College on a scholarship—many young men like him did.[26] John

Adams springs to mind. Even as it was, Franklin learned to read and, prior to his apprenticeship, received more formal education than Washington ever got. Several of his brothers prospered. That required ability and hard work, which Franklin supplied in large doses. Washington did so as well, with similar results. In colonial America, a new world of opportunities were open to the likes of Franklin and Washington.

Like Franklin's father, Josiah, Washington's father moved to Britain's American colonies from England—but the similarities end there. Josiah was a Puritan and left England after the Puritans ultimately lost the nation's religious civil war with the restoration of the Anglican monarchy in 1660 under the Catholic-leaning but hedonistic-living Charles II. Josiah naturally headed for the narrowly sectarian Puritan haven of Massachusetts. In contrast, Washington's father, Augustine, whose widowed mother had moved her young children back from Virginia to England with her second husband, returned around 1718 to claim his inheritance from his father's side. Fleeing England during the initial Puritan takeover that displaced his family, in 1656 Augustine's paternal grandfather had washed up in the loosely secular Anglican haven of Virginia. Once there, he began accumulating property on the frontier, including the plantation later named Mount Vernon, often by questionable means. The Natives, George Washington later wrote, called him "Caunotaucarius (in English) the Town taker."[27] Augustine's father continued the process of land acquisition and married well, providing substantial property for Augustine to reclaim when he reached maturity. He augmented it further.

"I was of a short-lived family," George Washington later noted of his male ancestors, with his father passing at age forty-eight when George was eleven.[28] Born on February 22, 1732, to Augustine's second wife, George had two older half brothers in an era when primogeniture assured that most inheritance went to the eldest son. The oldest sibling, Lawrence, received the Mount Vernon plantation while George's smaller share was controlled by his coarse, demanding, and selfish mother, Mary Ball Washington. Although of the planter class, George would have to work for his living and never received the formal education in England bestowed on his half brothers. With the hope of building his fortune in land speculation on the frontier like his forefathers, he opted to become a surveyor. Virginia still had a vast western wilderness. Surveyors were needed to inventory and divide it. Through their inside information, they could get in on the ground floor by acquiring some of the best parcels.

WASHINGTON FOUND HIS FIRST FORAY into the wilderness exhilarating. Six feet tall with broad hips and "blessed with a good constitution," as he later put it, he traveled in a surveying party with George William Fairfax, whose family held the largest proprietary land grant in Virginia.[29] A wedge-shaped slice encompassing more than five million acres between the Potomac and Rappahannock Rivers, the grant ran northwest from the Chesapeake Bay to the modern-day border of West Virginia and western Maryland. Mount Vernon lay in its already subdivided and settled eastern reaches, near the Fairfaxes' elegant seat of Belvoir Manor.

The party's goal was the northern Shenandoah Valley, a fertile part of the Fairfax Grant beyond the Blue Ridge not yet surveyed for sale. George William's father, William, was the grant's local land agent living at Belvoir; William's first cousin Thomas, an eccentric bachelor who held the title 6th Lord Fairfax of Cameron, was its current British proprietor and continually in need of revenue from it to maintain his lavish lifestyle. After Lawrence Washington hastily married William's fifteen-year-old daughter Ann in 1743, before a scandal broke involving her and a local Anglican priest, George Washington gained entry to Belvoir and be-friended George William Fairfax. First William and later Thomas took a liking to the young Washington, and became his mentors.

The expedition to the Shenandoah Valley in the spring of 1748 traversed a frontier region on the cusp of settlement. Washington, age sixteen, sometimes lodged in the crude cabins of early settlers but favored sleeping outside or in tents. His bed in one cabin, he wrote, was "nothing but a Little Straw—Matted together without Sheets or any thing else but only a Thread Bear blanket with double its Weight of Vermin such as Lice Fleas &c."[30] Plagued by heavy spring rains, the surveying party killed wild turkeys for food, swam its horses across swollen rivers, canoed down creeks, and at one point followed what Washington depicted as "the Worst Road that was ever trod by Man or Beast."[31] Judging by the words devoted to it in his journal, Washington's favorite experience involved a chance encounter with "thirty odd Indian coming from War with only one scalp." Washington went on: "We had some Liquor with us of which we gave them Part in elevating there Spirits put them in the Humour of Dauncing of whom we had a War Daunse." He branded the dance and accompanying music, which he described in some detail, as "a most comicle" spectacle.[32] Comfortable in the salons of Mount Vernon and Belvoir, Washington was also at home on the frontier and proved immune to its physical hazards.

Washington's work on this expedition must have impressed the Fairfaxes because he received ever more assignments to survey their lands, including in the Shenandoah Valley. Using income and insights from these jobs, Washington bought his first Shenandoah properties in 1750 and within two years had acquired more than two thousand acres there—a sizable holding for a youth of twenty who had had virtually no money of his own just a few years earlier.

Meanwhile, Lawrence Washington's health steadily deterio-
rated. He had contracted tuberculosis, which was then a common
disease with a shockingly high mortality rate. George accompanied him on various trips to obtain treatment, including to Bar-
bados in 1751, but nothing helped. He died in 1752, leaving Mount
Vernon to his young wife, Ann, and infant daughter Sarah, with
George next in line if they left no descendants in the Washington
line. When Sarah died two years later, Ann leased the plantation
to George on favorable terms. It passed to him when she died
in 1761. All in the family was the Washington way. Because of it,
George Washington suffered a morbid propensity to benefit from
family tragedy.

Lawrence's death also created a vacancy in the appointed position of Virginia adjutant general, the colony's top military post.
George wanted this too. Lawrence had earned the job as much
from his service leading Virginia troops in the 1741 Battle of Car-
tagena de Indias—an engagement during the War of Jenkins' Ear
between Britain and Spain, for which he obtained an officer's commission in the regular British army—as from his family connections and social status. Without any military experience, George
Washington had little to recommend him for the position other
than his relationship to the former occupant and support from the
Fairfax clan. That proved enough in colonial Virginia for its resident governor Robert Dinwiddie to split the post into four geo-
graphic divisions and for Washington, after some maneuvering, to
become district adjutant for the northern one, covering the Fairfax
Grant. Thus, at twenty, he leapfrogged to the rank of major in the
Virginia military. Now all he lacked to match his brother was a regular army officer's commission—a prize he long sought but never
received. America's most storied military career was launched
on the basis of family ties and a half brother's death rather than
proven ability.

The timing of this notable rise coincided with Washington's initial participation and rapid promotion within the Freemason movement. The first Masonic lodge in northern Virginia opened in 1752. Attracted (much like Franklin) to the ethical code and male fellowship offered by Freemasonry, Washington promptly joined it. He ascended to the rank of Master Mason within a year. Certainly in colonial Virginia, where the male Fairfaxes were active members, the Masonic movement provided the sort of connections that could advance a career. Because of the movement's secrecy, no one can now judge just how much membership helped Washington or Franklin. It surely helped some, and likely helped immensely.[33] For Franklin and Washington, Freemasonry provided an invisible bond that endured for life.

EUROPEAN CLAIMS TO LAND on the North American continent rested on exploration and occupation. In the case of Britain, its claims relied on John Cabot's sightings of the North Atlantic coast from Nova Scotia to Virginia in 1498, followed in time by the permanent occupation of the region, beginning with Jamestown in 1607 and Plymouth in 1620. Pennsylvania later filled in a key stretch in between. Britain's chief European rival, France, also made claims on the continent. It based them on the exploration of Canada, the Great Lakes region, and the Mississippi River south to the Gulf of Mexico by a series of explorers—from Jacques Cartier in 1534 through Samuel de Champlain in the 1610s to Robert de La Salle in 1682. Permanent French occupation began with Quebec in 1608 and extended to Louisiana by 1714. This pattern of expansion inevitably led to conflict as France's rapidly spreading but lightly settled fur-trading colonies reached around Britain's more slowly extending but more densely populated agrarian settlements. The first decisive clashes occurred in the North Atlantic region, where Britain wrested Acadia from France during Queen Anne's War in 1713.

By the 1750s, the tinderbox shifted to the Ohio River valley, where British settlers pushing west from Virginia and Pennsylvania butted against French efforts to occupy the space between its established provinces in Canada and Louisiana. Virginia based its claim to this so-called Ohio Country on its 1609 second charter, which purported to give it rights to all lands "from Sea to Sea, West and Northwest," and was interpreted by the mid-1700s to cover such lands at least as far west as the Mississippi River.[34] Under its 1681 charter, Pennsylvania received a superior claim to the land directly to its west as far as five degrees of longitude, which would later be found to include "the Forks of the Ohio"—present-day Pittsburgh—where the Allegheny and Monongahela Rivers join to form the Ohio. The first attempt by the British to occupy this region occurred in 1749, when the Virginia-based Ohio Company received a royal grant for 500,000 acres of land between the Fairfax Grant and the Ohio River. Maintaining the grant, however, required building a fort and settling at least one hundred families on the land within seven years. France countered by sending an expedition from Canada down the Ohio River to claim the region formally in 1749 and, four years later, by beginning to erect a line of forts along this route from Lake Erie toward the strategic Forks of the Ohio.

Upon learning of the French forts, Governor Dinwiddie, who had invested in the Ohio Company along with Lawrence Washington and other Virginians, asked his government in London how to respond. Instructions signed by the king directed Dinwiddie to send an emissary to the French commandant in the Ohio Country demanding an immediate withdrawal. If the French did not leave, the instructions to Dinwiddie added, "We do herby strictly charge, & command You, to drive them off by Force of Arms."[35]

Dinwiddie chose his newly minted adjutant for the northern district of Virginia, twenty-one-year-old George Washington, to deliver this message. Even Washington later acknowledged "the extraordinary circumstance that so young and inexperienced a Person should have been employed on a negotiation with which subjects of the greatest importance were involved."[36] On October 31, 1753, the same day that he received his orders at Virginia's capital, Williamsburg, Washington departed for the frontier. Assembling a seven-member party of interpreters and guides along the way that included the noted frontier scout Christopher Gist, Washington hoped to complete the trip before winter rendered the route impassible. The mission would make his reputation.

Crossing the Allegheny Mountains on horseback in driving fall rains and early winter snows, Washington's first objective was the Forks of the Ohio and a Native American meeting site near it, Logstown, where he hoped to recruit allies for his encounter with the French. At the time, despite the growing presence of British settlers and French fur traders, the Ohio Country remained under the effective control of various Native groups—particularly the Delaware, Mingo, and Shawnee—that sought to play each European power against the other for their own survival. Further complicating the political situation, the powerful, pro-British Iroquois Confederacy or "Six Nations," which was centered in the eastern Great Lakes region and represented locally by Tanacharison, whom the colonists called "Half King" for his status in the area, asserted a sort of vassalage over the weaker tribes of the Ohio Country.

Reaching Logstown after twenty-five days of hard travel, Washington recruited Tanacharison, two older tribal leaders, and a young Seneca hunter to go with his party to meet the French at their bases farther north. A larger Native contingent might raise suspicions, they explained, but in truth no others would go against the French. At this point in his journal, Washington related an eloquent speech by Tanacharison denouncing the French for invading the Ohio Country and establishing forts. Tanacharison praised "our Brothers the English" for not doing so. "The Land does not belong to one or the other," he said of the Europeans, "but the GREAT BEING above allow'd it to be a Place of residence for us."[37] Washington tactfully, some might say duplicitously, kept quiet about the Ohio Company's true intentions for the region. On his way to Logstown, he had spent two days scouting the Forks of the Ohio as a possible site for the Ohio Company's fort.

From Logstown, Washington rode overland with his expanded party some seventy miles north in five days to the fortified French trading post at Venango on the Allegheny River, then another sixty miles north in four days along the unfortunately named French Creek to Fort Le Boeuf, the southernmost French fort in the region. Snow, rain, and swollen rivers slowed their progress. Finally, on December 12, six weeks after he left Williamsburg, Washington met the French commandant for the Ohio Country and, a day later, delivered Dinwiddie's letter demanding that the French peaceably depart from "the Lands upon the River Ohio, in the Western Part of the Colony of Virginia." Naturally the French commandant refused: "As to the summons you send me to retire," he wrote back formally to Dinwiddie, "I do not think myself obliged to obey it."[38] As for English fur traders on the Ohio, whom the French had already begun to round up and send to Canada, Washington reported, "He told me the Country belonged to [the French], & that no English Man had a right to trade upon them Waters."[39] The forms of eighteenth-century European-style war had been satisfied; the fighting could begin.

More had happened than simply an exchange of letters. Observers apparently had shadowed Washington's party for weeks, with the same people turning up at different places. Charged with collecting intelligence on the French, at every stop from Logstown through Venango to Fort Le Boeuf, Washington heard similar reports from Native American informants, self-proclaimed French deserters, and French officers that thousands of soldiers were descending on the Ohio Country from Canada and Louisiana and dozens of local French forts were either built or planned. He saw only one fort and, by his own count, maybe a hundred men, but was assured that more than a thousand more soldiers had been temporarily withdrawn to winter quarters farther north. Also at every stop, the French tried to win over the Native leaders accompanying Washington by threats and bribes, warning them that the survival of their people depended on breaking from the British. "He was plotting every Scheme that the Devil & Man cou'd invent, to set our Indians at Variance with us," Washington said of the French commandant.[40] Treating him with every diplomatic courtesy, the French sent Washington back on December 16 with a story to tell of their resolve, belligerency, and might.

WHERE WASHINGTON'S OUTBOUND JOURNEY in late fall had been challenging, his midwinter return trip became a death-defying adventure. The horses were debilitated by the trek north, and Washington sent them ahead unloaded with three guides while he joined the rest in taking canoes down the ice-choked French Creek, a seven-day trip to Venango that he put at 130 miles due to all the river's bends and turns. "The ice was so hard we could not break through, but were obliged to haul our vessels across a point of land," Gist wrote at one point. "The creek being very low and we were forced to get out, to keep the canoes from oversetting, several times; the water freezing to our clothes," he noted at another.[41]

Washington planned to remount at Venango for the overland trek to the Forks of the Ohio, but the horses still proved "so weak

& feeble," as he put it, that the Virginians took up their gear in packs and walked.[42] With the formalities finished and a hard march ahead, Washington discarded his military uniform in favor of leather leggings and a blanketlike matchcoat—an "Indian walking Dress," he called it.[43] Tanacharison and the two other Native leaders remained behind at Venango.

"I can't say that ever in my Life I suffer'd so much Anxiety as I did in this affair," Washington lamented at the time about the French. "I saw that every Stratagem that the most fruitful Brain cou'd invent: was practiced to get Half King won to their interest, & that leaving of him here, was giving them the Opportunity they aimed at."[44]

Begun as a group effort with lame horses, the overland trek proved unnervingly slow to Washington. "The Horses grew less able to travel every Day," he complained. "The Cold increas'd very fast, & the Roads were getting much worse by a deep Snow continually freezing."[45] In haste to reach Virginia with his report, after three days of slogging through the forest with the others, which included a very bleak Christmas, Washington left the group behind and, accompanied only by Gist, rushed on toward the Forks of the Ohio and home.

"The Day following," Washington wrote, "we fell in with a Party of French Indians, which had laid in wait for us."[46] Gist remembered it as only a single male, whom he had seen with the French at Venango and now offered to serve as their guide. At some point, both Washington and Gist noted, the Native turned and fired toward them at short range—a mere fifteen feet, according to Washington. "Are you shot?" Washington asked Gist. "No," Gist replied.[47] Then they both set upon their assailant. Gist wanted to kill him. Washington instead held him until dark and then, pretending to make camp, set him free. Once the assailant was some distance away but before he could return with others, the Virginians made a dash toward the Allegheny River, walking all night and the entire next day, until they felt safe enough to camp before crossing the river above the Forks of the Ohio, where they hoped to collect fresh horses for the trip home. This was the first of many such brushes with death at close quarters for Washington—experiences that led some to see his life as charmed.

Reaching the Allegheny on December 29, Washington hoped to find it frozen for the crossing. Instead, the middle churned with rushing water and broken ice. With Gist, he spent the day building a makeshift raft. Launched after sunset in a waterscape that must have looked like Emanuel Leutze's 1851 painting of Washington crossing the Delaware on Christmas night twenty-three years later, the raft did not make it far across this raging river. "Before we got half over," Washington wrote, "we were jammed in the Ice in such a Manner, that we expected every Moment our Raft wou'd sink, & we Perish." Using a crude punting pole to stop the raft short of a menacingly large chunk of oncoming ice, he added, the rushing floe hit with "so much Violence against the Pole, that it Jirk'd me into 10 Feet Water."[48] A strong swimmer but perilously at risk in the frigid water, Washington grabbed onto the raft and got back on board but by then it was useless to try to guide the raft toward either shore. Gist had frostbite and Washington was drenched and in danger of hypothermia. As their raft carried them past an island in midstream, the men jumped off and waded ashore. "We contented ourselves to encamp upon that island," Gist dryly noted.[49] By morning, the extreme cold had frozen the river's surface so completely that the men could walk safely to the far shore and on to the Forks of the Ohio.

Having reached familiar territory, one final ominous episode awaited Washington and Gist. At the Forks they met a party of friendly Natives who had recently come across the scene of the massacre by pro-French Ottawas of a British settler's family, with the parents and five children scalped and their bodies left to be eaten by their hogs. It was the sort of tragedy that had often occurred on the frontier during the last war between the French and the British, and would become common again during the coming one.

Within a year, the western reaches of Pennsylvania and Virginia would stand at the center of widening war. Their citizens would turn to Franklin and Washington for leadership and, by doing so, bring together these two men—the former rich in experience; the latter brimming with youthful ardor. "Washington was one of the few prominent members of America's founding generation—Benjamin Franklin was another—who were born early enough to develop their basic convictions about America's role in the British Empire within the context of the French and Indian War," historian Joseph Ellis observed.[50] It gave them a shared vision that shaped their approach toward colonialism, independence, and American nationalism.

Racing from the Forks of the Ohio to Williamsburg in two weeks, Washington delivered the terse French response and his own detailed report to Governor Dinwiddie on January 16, 1754. Recognizing the propaganda value of Washington's account, Dinwiddie immediately ordered it typeset and published for wide distribution. Nothing less than a clarion call for war, the hastily written *Journal of Major George Washington* was widely serialized in colonial newspapers and reprinted as a small book in London, with its poor spelling, crude grammar, and blunt style simply adding to its credibility and sense of urgency. The narrative value alone assured its popularity, but the theme of French belligerence and treachery made it useful for stirring colonists for the inevitable war ahead. "On his veracity the most cogent arguments for Colonial military action were based," biographer James

Thomas Flexner wrote of Washington. "He was the physical embodiment of the war party."[51] Almost overnight, this publication transformed the young militia officer into a frontier hero—a reputation that the coming French and Indian War would further burnish.

THE PUBLISHED VERSION of Washington's *Journal* reached Philadelphia by March 1754, and found an interested reader in Franklin. Of course, Franklin had led local efforts to defend Pennsylvania from the French and their Native American allies during King George's War in 1747. Three years later, he wrote "Observations Concerning the Increase of Mankind," an influential essay which argued that, because of abundant cheap land on the frontier and the tendency for people to marry young and have many children when economic conditions permit, the population of Britain's American colonies would double every two decades and, simply by natural increase, exceed that of England within a century—which it did. Although this essay celebrated the expansion of America within the British Empire, some passages suggested that growth might lead to independence and several admonished Britain to defend and extend the frontier. "How careful should she be to secure Room enough, since on the Room depends so much the Increase of her People," Franklin wrote of the motherland. "So vast is the Territory of North-America, that it will require many Ages to settle it fully."[52]

Furthermore, in September 1753, when French activities in the Ohio Country increased with the building of the forts that sent Washington west later that same year, Pennsylvania appointed Franklin as one of three commissioners sent to negotiate with Native American leaders from the region toward maintaining or gaining their allegiance. Meeting with representatives of the friendly Iroquois Confederacy, the wavering Delaware, and other Ohio Valley tribal groups at the frontier town of Carlisle, the commissioners provided material support, including guns, and won a tentative pledge of support. Franklin wanted to do more for the Natives and left the four-day conference deeply committed to the defense of the frontier at a time when Pennsylvania's self-interested proprietors and pacifist Quaker officials remained reluctant to act decisively. He later published his transcriptions of the sum of the speeches given at this and other treaty negotiations, creating an important record of Native American oratory.

A gifted advocate for any cause that he favored, Franklin saw an ally in Washington and recognized the value of his *Journal* in alerting Pennsylvanians to the threat posed by the French. Franklin clearly endorsed the Virginian's efforts and embraced his call to arms. Franklin's *Pennsylvania Gazette* soon began running stories based on Washington's *Journal*, including the French commandant's belligerent response to Dinwiddie's letter, accounts of the massacre of British settlers by pro-French Natives, and Washington's reports of both the French military buildup in the Ohio Country and French efforts to seduce "our Indians" to their side.[53] Based on his reputation as a military organizer gained in the prior war with France, his leadership position in the colonial assembly, and his influence as a writer and publisher, Franklin became as much the physical embodiment of the war party in Pennsylvania as Washington was in Virginia. The coming French and Indian War would reinforce the political stature of both men. At first fighting for Empire, their cause became America.

Two

LESSONS FROM THE FRONTIER

WASHINGTON'S REPORT FROM THE FRONTIER that France had launched a military occupation of the Ohio Country and refused to withdraw prompted immediate responses from Williamsburg, Philadelphia, and London. These responses pulled Franklin and Washington into converging efforts by then adjacent colonies with overlapping claims to the frontier.

With the Ohio Company already scrambling to fortify the Forks of the Ohio, Virginia formed a regiment, promoted Washington to the rank of lieutenant colonel, and sent him with two thin companies of raw recruits as the vanguard of a larger force to defend the frontier. Having previously attempted to reinforce and extend its alliances with the Native peoples of the Ohio Country through negotiations and payments at Carlisle, Pennsylvania began pushing a plan formulated by Franklin to forge a coordinated response to the French by establishing a formal intercolonial union. With all its American colonies imperiled by the rising French threat in the Ohio Country, Britain dispatched two regular army regiments from Ireland under the command of General Edward Braddock to drive the French out of the region once and for all. With both men drawn into all three of these responses, the lives of Franklin and Washington began intertwining.

ON MAY 28, 1754, Washington's green Virginia troops drew first blood in what would become the global Seven Years' War between the far-flung British and French Empires. The war lasted nine years in North America, where British colonists called it the French and Indian War. They had never seen such carnage.

It all began some fifty miles south of the Forks of the Ohio. Approaching that point in late May with about 160 armed men and light artillery, Washington heard that "a Body of One Thousand *French* and upward"—the actual count was perhaps half that number—had driven off thirty or so Ohio Company agents engaged in building a crude fort at the Forks of the Ohio and begun constructing more formidable fortifications of their own, Fort Duquesne.[1] This was to become the anchor of French interest in the Ohio Country and the linchpin in a chain of forts linking Canada and Louisiana.

With ill-advised bravado, Washington led his small band relentlessly forward, cutting a crude road over the forested mountains dividing the Potomac and Monongahela River valleys as a route for further troops, supply wagons, and cannons to mount an assault on Fort Duquesne. The road would run from the confluence of Wills Creek with the North Branch of the Potomac, at the present site of Cumberland, Maryland, to the place thirty-eight miles south of modern-day Pittsburgh where Redstone Creek joins the Monongahela. Driving rains swelled the rivers and slowed progress. Some days they covered only a few miles; other days none at all.

Washington received repeated reports from Native informants and retreating British traders of the mounting French buildup ahead and of forces sent out to halt his advance. He sent regular dispatches back to Governor Dinwiddie calling for more men and sent letters to the leaders of Pennsylvania and Maryland alerting them that France had taken the Forks of the Ohio and asking for their colony's help to recover it. "I have heard of your Honour's great zeal for His Majestys Service; and for all our Interest's on the present occasion," he implored the Pennsylvania governor, and he admonished Maryland's executive that French aggression in Ohio "should rouse from the lethargy we have fallen into, the heroick spirit of every free-born Englishman."[2] Washington also wrote ahead to the Iroquois Half King Tanacharison and other Native leaders calling on their support, only later to discover that some were shifting sides in response to French promises and threats.

Little immediate help came from either direction, even though Dinwiddie pledged that reinforcements were on the way under the command of Colonel Joshua Fry, whom the governor had entrusted with overall command of the Virginia Regiment. A small force of South Carolina regulars was also coming but, even without them, Washington opted to press on with road building in the hope of reaching Redstone Creek, where the Monongahela River offered direct water carriage to the Forks of the Ohio. His recruits, Washington complained, generally "are of those loose, Idle Persons that are quite destitute of House, and Home, and, I may truely say, many of them of Cloaths." Some could shoot but most lacked discipline. "In short," he wrote, "they are as illy provided as can well be conciv'd."[3]

Crossing over the main ridge of mountains and into what became southwestern Pennsylvania, the Virginians reached a large clearing in the forest known as Great Meadows. After removing some brush and improving two natural entrenchments, Washington depicted it as "a charming field for an Encounter" and began sending out scouting parties looking for enemy forces.[4] On May 27, the trusted guide Christopher Gist brought word that fifty French soldiers had visited his nearby farm a day earlier, prompting Washington to dispatch half his men in hot pursuit. Toward

evening, a messenger arrived from Tanacharison stating that he
had followed footprints to a rocky hollow or glen where a small
French force had taken shelter.

Washington divided his remaining men, leaving some to guard
the supplies while he led forty others on a rainy, nighttime trek
though the pitch-black forest to meet Tanacharison and a dozen
Native warriors. Acting in consort, the colonial and Native fight-
ers surrounded the French position and racked it with fire from
above. Accounts vary as to which side fired first, but the Virginians quickly got the better of it. With ten on their side dead within
fifteen minutes, including their young leader Joseph Coulton de
Villiers, sieur de Jumonville, the twenty-one surviving French soldiers surrendered, claiming they were merely emissaries sent to
convey a summons directing the Virginians to leave French territory. The two nations were not yet at war.

Branding Washington as an assassin, the French asserted that
de Jumonville was shot in cold blood while trying to deliver that
summons. Other reports had Tanacharison splitting the officer's
skull with a tomahawk and scooping out the brains. Various accounts agree that Native warriors scalped all the fallen French-
men, with Virginian reports adding that Washington stopped
them from scalping the survivors. He sent those survivors as prisoners of war to Virginia, where he was again hailed as a hero. In
a letter to his brother John, Washington wrote of standing in the
line of enemy fire: "I heard Bulletts whistle and believe me there
is something charming in the sound."[5]

The reaction was more muted in Britain. Dinwiddie had orders
to reclaim the Ohio Country without precipitating a war. The
French should have an opportunity to withdraw peacefully, which
Washington had not given them. After reading Washington's letter in a London magazine, a battle-hardened King George II commented that the young Virginian might not find the whistle of

bullets so charming if he heard it more often. The rebuke was clear but also clearly wrong. While Washington may have hated war, he was always drawn to warfare. And he vehemently denied the assertion that de Jumonville and his men were emissaries. "Instead of coming as an Embassador, publickly, and in an open Manner," Washington noted in his journal, "they came secretly, and sought after the most hidden Retreats, more like Deserters than Embassadors." He branded their claim to have called out their status before taking up arms as "an absolute Falshood."[6]

THE FRENCH COMMANDER at Fort Duquesne was de Jumonville's older brother. Seeking revenge, he reacted swiftly by sending an overwhelming force of Canadian troops and Native warriors against the Virginians. Both Dinwiddie and Tanacharison urged Washington to retreat, but he instead merely fell back to Great Meadows, where he ordered his men to build a split-log stockade with supplementary low breastworks and trenches. Dismissed by Tanacharison as "that little thing upon the meadow," Washington called it Fort Necessity.[7] Although in an expanse of open space, the surrounding forested hills were close enough to shield snipers and heavy summer rains could quickly turn trenches into mudholes. While even determined attackers probably could not storm the fort against armed defenders, those defenders could not rush out against attackers, who were likely capable of picking them off one by one with musket fire from protected positions. In short, Fort Necessity was a foreseeable death trap against a large, well-armed force. Washington expressed confidence, however. "We expect every Hour to be attacked by superior Force, but shall if they forbear one day longer be prepared for them," Washington wrote in his May 31 letter to his brother John. "We have already got Intrenchments & are about a Pallisado'd Fort, which I hope will be finished today."[8]

Reinforcements trickled in from the east. Some two hundred raw Virginia recruits arrived without their commander, Joshua Fry, who had died in a fall from his horse. With this chance occurrence, one of so many that advanced his career, Washington assumed overall command of Virginia's frontier troops with the rank of colonel.

Another hundred soldiers came from South Carolina under a Scottish officer, James Mackay, who by virtue of his king's commission outranked Washington and maintained independent authority over his troops. Rather than dig in for the coming fight, Washington directed his men to resume road building toward Redstone Creek with the ultimate object of capturing Fort Duquesne and driving out the French. The French instead used that road to attack Fort Necessity.

In late June, as a force reported to include some eight hundred Canadians and four hundred Natives approached, the colonials fell back to Fort Necessity and their own Native allies fled. By this point, Tanacharison considered Washington rash and the campaign hopeless. The assault began shortly before noon on July 3, scarcely five weeks after the bloodletting at what became known as Jumonville Glen. Although the actual size of the enemy force was not as large as first reported, it was more than adequate to rain down death on the fort from protected positions in the surrounding hills and forest. Washington first tried to fight in the open outside the fort, but when heavy musket fire began, his undisciplined Virginians ran for cover, forcing Washington and the South Carolinians to follow. Fort Necessity offered limited protection from bullets, especially after a midafternoon rainstorm flooded its trenches and drenched stored gunpowder. By nightfall, a third of the defenders were dead or wounded, many of the survivors drunk, and all their horses and cattle slaughtered by enemy fire. When the French commander offered reasonable terms for capitulation, Washington and Mackay had little choice but to accept them.

In line with French claims to the Ohio Country and in the absence of a formal state of war between Britain and France, the commander simply demanded that the British troops withdraw across the mountains and return those captured in the prior battle. They could take what they could carry, including their insignia and flags. The first sentence of the articles of capitulation, written in French by the French commander, stated the essence of the agreement. "Our Intentions have never been to trouble the Peace and good Harmony subsisting between the two Princes in Amity," it declared, "but only to revenge the Assassination committed on one of our Officers, bearer of a Summons, as also on his Escorte."[9] Not knowing French, Washington later claimed that he did not understand the meaning of the French word "l'assassinat" or that, by signing the document, he confessed to a dark and dishonorable act.[10] Yet the word clearly appeared in the document and its meaning was plain to anyone who read French. French propagandists made the most of it in the run-up to the Seven Years' War, just as British propagandists would during the American Revolution.

On July 4, 1754, twenty-two years before that day in July became associated with American independence, Washington, Mackay, and their remaining men marched out of Fort Necessity and into an uncertain future. War was coming, they certainly knew, and British prospects in the Ohio Country looked bleak so long as the colonies were divided and lacked sufficient support from Britain. As news of the capitulation spread, others realized this as well, including Franklin (who trumpeted the story in his *Gazette*) and the government in London. Although harshly criticized in Canada and France for his actions at Jumonville Glen—"There is nothing more unworthy and lower, and even blacker, than the sentiments and the way of thinking of this Washington," the governor-general in Quebec wrote[11]—the young colonel's reputation for bravery on the battlefield remained undiminished in Virginia. Along with Franklin, he would play a key, ongoing role in ensuing events.

FRANKLIN SAW WASHINGTON'S ON-THE-GROUND REPORTS from the frontier as proof that the thirteen colonies needed to join in common cause. Indeed, no sooner had Franklin learned from Washington's letter to the Pennsylvania governor that the French had taken the Forks of the Ohio than he began shopping around an earlier idea for an intercolonial defensive confederation—a first step toward the sort of union that he would later work with Washington to craft. Their purpose, Franklin editorialized in the *Pennsylvania Gazette* about the French, is "to establish themselves, settle their Indians, and build Forts just to the Back of our Settlements in all our Colonies, from which Forts . . . they may send out their Parties to kill and scalp the Inhabitants, and ruin the Frontier Counties." British settlers were already fleeing the frontier, Franklin wrote, and British traders being captured along with their goods. The settled regions were next, he warned.

"The confidence of the French in this Undertaking seems well grounded on the present disunited State of the British Colonies, and the extreme Difficulty of bringing so many different Governments and Assemblies to agree on any speedy and effectual Measures for our common Defense and Security," Franklin declared. Beneath this appeal he printed the first original editorial cartoon in any American newspaper, and perhaps the most famous ever published. Presumably of his design, it showed a rattlesnake cut into pieces with the name of a colony on each severed part. "Join, or Die," the caption read.[12] Fittingly, he chose a native American species for the illustration, not an introduced

European domestic animal. United, the image suggested, America could survive on its own.

Franklin's promotion of an intercolonial union received a boost later that year when, based on Washington's prior report of French aggression, the Board of Trade in London directed the colonies to send commissioners to a congress in Albany, New York, to consider collective defensive action and to mend their military alliance with the Iroquois Confederacy. Instinctively viewing collective action as the way to solve social and political problems, Franklin had sketched out an idea for his intercolonial defensive union in a 1751 letter to James Parker, a New York printer. As Franklin then proposed it, this limited federal union, with proportional representation from the various colonies and a governor appointed by the king, would manage "every Thing relating to Indian Affairs and the Defence of the Colonies." Using the Iroquois as an example, Franklin asked, "If six Nations of ignorant Savages should be capable of forming a Scheme for such an Union, and be able to execute it," why should a like union "be impracticable for ten or a Dozen English Colonies, to whom it is more necessary, and must be more advantageous?"[13]

Chosen as one of Pennsylvania's commissioners to the Albany Congress, Franklin dusted off his earlier idea for this defensive union and drafted it into a formal proposal. Only seven middle and northern colonies sent commissioners to the congress, and most of these directed their delegates to focus on negotiations with the Iroquois. Nevertheless, working with Thomas Hutcheson of Massachusetts, Franklin persuaded the congress to appoint a committee to draw up a plan of union and then persuaded that committee to adopt his proposal. While each colony would retain authority over its internal affairs, a "General Government" with a president general appointed by the king and a grand council composed of representatives from the various colonies would control relations with Native Americans, settlement of the west, and defense of the coast and frontier. For these purposes, it could raise an army, launch a navy, make laws, and levy taxes.

Although ahead of its time, the Albany Plan of Union contained seeds of the novel sort of federal government that Franklin would help to design three decades later at the Constitutional Convention. "Though he was sometimes dismissed as more a practitioner than a visionary," biographer Walter Isaacson later wrote, "Franklin in Albany had helped to devise a federal concept—orderly, balanced, and enlightened—that would eventually form the basis for a unified American nation."[14] He never gave up on this vision. He carried it to the Constitutional Convention.

While it was approved by the Albany Congress and passed on to the colonies and the British government for implementation, Franklin's plan never had a chance. No colony was yet willing to relinquish its sovereignty over military affairs and the frontier. Parliament feared that *any* intercolonial union might lead toward independence.

To gain British support for the Albany Plan, the royal governor of Massachusetts suggested that the king (rather than the colonial assemblies) choose the grand council much as he chose the upper house of some colonial legislatures. Again showing foresight, Franklin rejected this suggestion on principle. "It is suppos'd an *undoubted Right of Englishmen* not to be taxed but by their own Consent given thro' their Representatives," he declared in an early articulation of the revolutionary rallying cry: No taxation without representation.[15]

"Its Fate was singular," Franklin observed about the Albany Plan. "The Assemblies did not adopt it as they all thought there was too much *Prerogative* in it; and in England it was judg'd to have too much of the *Democratic*."[16] Franklin remained convinced that, if implemented, the plan would have sufficed to defend the colonies against the French, but instead Britain was forced to intervene at such great cost that it led to the taxes that touched off a revolution. "But such mistakes are not new," Franklin later sighed. "History is full of the Errors of States and Princes."[17]

FOLLOWING THE FAILURE of Washington's military assault and Franklin's defensive union, Britain sent a fourteen-hundred-man army of crack British soldiers under the command of General Braddock—two infantry regiments of the Coldstream Guards—to drive the French from the Ohio Country. This was a far larger force than anything the French in Canada could muster and should have been more than sufficient to succeed. The Ohio Country had never seen such a sizable, trained, European army before; indeed, nothing like it had ever marched so far into the region's interior. An imperious leader dismissive of colonists, Braddock assumed his army would continue past the Forks of the Ohio to Lake Ontario, rolling up a line of French forts on its way. For this purpose, it traveled with more than two dozen heavy siege cannons, howitzers, and mortars in addition to a long, cumbersome supply train. At the western end of Lake Ontario, Braddock's force was to link up with a second one marching west from Albany, thus completing the conquest of the Ohio Country.

Reaching Virginia in February 1755, Braddock met with Franklin, Washington, and other colonial officials over the next two months while finalizing his plan of attack. Pennsylvanians hoped that he would march on Fort Duquesne through their colony, which was the shortest way, while Virginians urged him to follow Washington's route. The latter option would allow Braddock to take advantage of both Fort Cumberland, which Maryland had just built at the Wills Creek jumping-off point from the Potomac to the Monongahela River valley, and Washington's crude road from that point to Great Meadows. The general used Washington's route and recruited the Virginian to serve as an aide and guide. In his role as postmaster general for the colonies, Franklin met with Braddock to discuss communications and ended up serving as a key supply agent as well.

Their work with Braddock brought Franklin and Washington together for the first time, possibly at Frederick, Maryland, in April. By this time, each man knew the other by reputation. Franklin had published excerpts of Washington's letters and journal from the frontier in the *Pennsylvania Gazette*. An avid consumer of London magazines, Washington must have read about Franklin, who was by then the most famous colonist in the British Empire.

Franklin arrived at Frederick with his son William on April 18 to set up express mail routes to facilitate communications between the army and key colonial capitals. There he learned the extent of Braddock's anger at Pennsylvania for not providing financial support for the British troops and at Maryland and Virginia for not supplying 150 promised wagons and four-horse teams to transport the army's supplies across the mountains from Fort Cumberland to Fort Duquesne. Pennsylvania's governor, a lackey of the colony's British proprietors, had blamed the lack of support on the colonial assembly's pacifist Quaker party. A leading non-Quaker member of that assembly, Franklin explained to Braddock that the legislature had voted the funds, only to have the governor veto the bill because it did not give him control over disbursements. As for wagons and horses, Franklin offered to procure them from Pennsylvania farmers on fair terms. Braddock was elated and, when Franklin delivered on his promise, said it was "almost the only Instance of Ability and Honesty I have known in these Provinces."[18]

Franklin secured the wagons and horses by printing a public advertisement offering "fair and equitable" rental terms for a projected four-month-long campaign, but with a devious twist.

Braddock was "extremely exasperated" at not having received the needed wagons from Pennsylvania and, if they were not delivered by May 20, would likely send "Sir John St. Clair, the Hussar, with a Body of Soldiers" to secure them, Franklin warned in a concluding reference to the general's abusive quartermaster: "Violent Measures will probably be used." As a "Friend and Well-wisher" who had "no particular Interest in this Affair," Franklin advised the farmers to accept the offer and "make it easy to yourselves."[19] Of course, Braddock expected the wagons from Maryland and Virginia, not Pennsylvania, and never threatened to commandeer them. Although St. Clair had roughly treated workers engaged in building local supply routes, he was neither a hussar mercenary nor under orders to confiscate wagons. Still, the ploy worked and Franklin was widely praised for securing the wagons and horses. "I cannot but honour Franklin for the last Clause of his Advertisement," Braddock's military secretary wrote to the Pennsylvania governor.[20]

Braddock dispatched letters to his superiors in London commending Franklin. The two men, the former sixty and the latter nearly fifty, got along famously and dined together often. "This general was I think a brave Man, and might probably have made a Figure as a good Officer in some European War," Franklin later commented. But when he warned the general about the risk of ambush by Native warriors on the long march to Fort Duquesne, he recalled Braddock smugly replying, "These Savage may indeed be a formidable Enemy to your raw American Militia; but upon the King's regular and disciplin'd Troops, Sir, it is impossible they should make an Impression."[21] Washington would later give Braddock a similar warning, to the same effect. Before Franklin left Frederick, he also arranged for gifts of supplies, including coffee, chocolate, and rum, for the army's junior officers, winning more goodwill for himself and Pennsylvania.

AS FRANKLIN WAS CONCLUDING THESE ARRANGEMENTS, Washington was rushing toward Braddock's headquarters in Frederick. He had experienced a life-altering nine months since his capitulation at Fort Necessity. First, he came to terms with his own defeat by recognizing his rashness and learning from it. Second, as shown by his rising social status, although Washington was still admired in Virginia for his military exploits in part due to wildly exaggerated claims of enemy casualties at Fort Necessity—some three hundred rather than the actual three—Dinwiddie turned against his former protégé in light of British displeasure with the young officer's failures. After the loss of men at Fort Necessity, the governor re-formed the Virginia Regiment into local companies with a mere captain in charge of each. Not willing to accept demotion, Washington resigned. Third, Washington became master of Mount Vernon and its eighteen slaves following the death of his eldest brother Lawrence's last surviving child, Sarah, at age four in 1754. Technically, he rented the family plantation from Lawrence's remarried widow until her death in 1761, but, under Lawrence's will, he would then inherit it. Born the third son and by a second wife, Washington succeeded to the status of landed gentry though the death of Lawrence and all his heirs.

Finally, once Braddock had arrived in February 1755, with two regiments charged with the capture of Fort Duquesne, Washington became invaluable as the frontier military guide who best knew the terrain ahead. In warfare, experts affirm, geography is destiny. Through his chief aide Robert Orme, Braddock invited Washington onto his personal staff.

For his part, Washington welcomed the chance to serve in a real army under a respected general with the prospect of a royal commission as a regular British officer. "My inclinations are strongly bent to arms," Washington wrote late in 1754.[22] At least initially, Braddock could offer nothing higher than a temporary captaincy, but with victory, more was possible, and so Washington volunteered to serve as an aide-de-camp with nothing beyond his rank as a former colonial officer and the hope of a future British colonelcy. In April 1755, he was dashing to catch up with Braddock at Frederick in anticipation of a triumphant march on Fort Duquesne and revenge for his defeat at Fort Necessity.

By the end of May, Braddock and his British regulars, Washington and companies from four different colonies, and Franklin's wagons—about 2200 men in all, along with some 120 camp women serving in domestic roles—converged on Fort Cumberland to begin the 110-mile march over rugged, densely forested mountains toward Fort Duquesne. The main missing components were Native warriors and scouts, nearly all of whom Braddock had alienated by refusing to recognize their right to permanently remain in the Ohio Country.

The road proved far worse than expected, forcing the army to grade and widen it along the way so that the artillery and heavy wagons could pass. Departing Fort Cumberland in early June, the troops moved slowly until, at Washington's suggestion, Braddock split his force between an advance column of about thirteen hundred proven soldiers and a support column that trailed ever farther behind improving the road and bringing up most of the heavy artillery and supplies.

By this point, Washington had fallen gravely ill with violent fevers and headaches, likely from the dysentery sweeping through camp, and severe hemorrhoids, which forced him at first to ride in a wagon with the advance column and then, upon Braddock's orders, to remain behind until sufficiently recovered to travel. Any other course, the doctors warned, could prove fatal. The only concession that Washington could wrest from Braddock was a promise to call him forward before the assault on the fort. Thanks to patent medicine suggested by Braddock, Washington felt himself "tolerably well recoverd" by July 2 and, again traveling by wagon, caught up to Braddock's main force on July 8, some twelve miles from Fort Duquesne.[23]

Up to this point, the British had encountered surprisingly little resistance from the French or their Native allies, who closely monitored Braddock's advance. In the dense, unfamiliar forest, however, waiting for an attack could be as terrorizing as an actual one, especially for soldiers conversant in stories about the stealth and savagery of Native Americans. In reality, Native warriors preferred taking able-bodied captives, whom they could parade as trophies, use as slaves, and ransom as hostages. Scalping enemies produced a lesser prize, but Braddock's men brooded more on the latter fate, which typically befell the dead and wounded. Neither diabolical nor depraved, Native Americans acted sensibly in light of their objectives and values, a truth that Franklin understood better than Washington.

Lacking sufficient men of his own, Captain Claude-Pierre Pécaudey, seigneur de Contrecoeur, the commander at Fort Duquesne, had recruited an army of Native warriors to his side, mostly Ottawa, Wyandot, Potawatomi, and Ojibwe from the Great Lakes region. These warriors, he knew, would fight, not to defend the fort, but to kill and capture British soldiers. With the British within a day's march of Fort Duquesne, on the morning of July 9, Contrecoeur dispatched more than 600 Native warriors, along with 250 French officers, colonial regulars, and Canadian militiamen—roughly half of his entire force but less than two-thirds of Braddock's advance column—to strike the British before they reached the fort. The British set off that same morning from their camp about ten miles south of Fort Duquesne in a mile-long column. A forward unit of 300 British regulars led the way, followed by a company of New York soldiers guarding 250 civilian road workers. The main body of 500 British infantry marched in long, parallel lines flanking supply wagons, artillery, and camp women. With cushions strapped to his saddle to soften the ride for his hemorrhoids, Washington rode alongside Braddock in the main force. A rear guard of Virginians trailed behind while small parties patrolled the flanks.

Around 1:00 P.M., after fording the Monongahela River for a second time that morning, the British entered an open woods cleared of underbrush as a Native hunting ground. Trees offered cover for hunters without obstructing their line of fire. By all accounts, the British were tired from their march and anxious in their surroundings while the Natives were fresh and at home in the forest. Here, the two armies met without warning.

Three quick volleys by the British forward unit disrupted the French regulars but sent the Natives streaming into the woods on either side of the British column, which they rained with fire from protected positions. Easy targets, mounted officers fell first: fifteen of the forward unit's eighteen within ten minutes. Leaderless, this unit fell back just as the main force rushed forward and the unarmed road workers dashed for cover, causing pandemonium in the ranks and added losses by friendly fire.

Trained to stand together and fire at similarly ordered enemy soldiers, the British found themselves exposed to a withering barrage with no obvious targets for return fire. "The French and Indians crept about in small Parties," one survivor wrote, "and in all the Time I never saw one, nor could I on Enquiry find any one who saw ten together. The Loss killed and wounded 864."[24] With four horses shot out from under him, Braddock tried to rally his men until he dropped to the ground with a bullet in his chest—a mortal but not immediately fatal wound. Then the survivors retreated in an ever more panicked state across the river to their rear, leaving most of their wounded, baggage, wagons, and supplies. Nearly all the camp women were killed or captured.

Washington, Braddock's only uninjured aide-de-camp and remarkably calm under fire, loaded the wounded but still conscious
general into a cart, escorted him across the river, and then rode
back to form a rear guard and help restore order to the fleeing
troops. His action saved many. "I have been protected beyond all
human probability & expectation," Washington later reported to
his brother, "for I had 4 Bullets through my Coat, and two Horses
shot under me yet although death was levelling my companions
on every side of me, escaped unhurt."[25]

Nearly two-thirds of Braddock's officers and staff were killed
or wounded during the three-hour battle. The ratio was similar
for enlisted men. One credible count put total British casualties at
976 out of 1469, compared with only about 40 for the other side.
Of all the British units engaged in combat, only the Virginian rear
guard fought effectively by breaking rank and shooting from behind trees, even though this tactic led some British regulars to
mistake the crouching Virginians for the enemy and fire at them.
"You might as well send a Cow in pursuit of a Hare as an English
Soldier . . . after Canadeans in their Shirts, who can shoot and
run well, or Naked Indians accustomed to the Woods," the rear
guard's captain complained afterward.[26] "The Virginians behavd
like Men, and died like Soldier's," Washington advised Governor
Dinwiddie, adding that out of about 150 Virginians, "scare 30
were left alive."[27]

Nothing but total exhaustion could stop the flight back toward
the support column, then some fifty miles behind. They "broke
& run as Sheep before Hounds," Washington wrote to Dinwiddie
about the British regulars, "and when we endeavored to rally them
in hopes of regaining the ground and what we had left upon it, it
was with as little success as if we had attempted to have stopd the
wild Bears of the Mountains."[28] Gleaning the battlefield for treasure
and trophies, the Natives had no intention of pursuing the fleeing

troops, but the terrified survivors had no way of knowing this. Unable to rally them, Washington returned to Braddock, who ordered him to ride through the night to the support column and instruct its commander, Thomas Dunbar, to cover the retreat. "The shocking Scenes which presented themselves in this Nights march are not to be described. The dead—the dying—the groans—lamentation— and crys along the Road of the wounded for help," Washington commented, "were enough to pierce a heart of adamant."[29]

Two more days brought most of the able-bodied survivors back to Dunbar's camp. Aside from Braddock, who arrived in a cart, virtually all of those unable to walk were left to die or be killed. The combined British force still vastly outnumbered the French at Fort Duquesne, especially after most of the Native warriors left with their plunder, but neither Braddock nor Dunbar gave any thought to resuming the attack. They ordered their men back to Fort Cumberland. Braddock made it only as far as Great Meadows, where he died from his wounds. His body was interred under the roadway in an unmarked grave that the troops marched over so that no one could find and molest the body. However much Washington disparaged the British soldiers, the defeat resulted from bad leadership, both strategic and tactical, for which he, as a mere aide-de-camp, bore little blame. Indeed, Washington emerged from the debacle with an enhanced reputation for courage and survival, if not success, on the battlefield.

"I Tremble at the consequences that this defeat may have upon our back settlers," Washington wrote to Dinwiddie from Fort Cumberland on July 18. "Colo. Dunbar, who commands at present, intends as soon as his Men are recruited at this place, to continue his March to Philia[delphia] for *Winter* Quarter's; consequently there will be no Men left here unless it is the shattered remains of the Virginia Troops; who are totally inadequate to the protection of the Frontiers."[30]

Washington did not exaggerate the danger. Upon hearing of the defeat, Franklin warned, "We have lost a Number of brave Men, and all our Credit with the Indians; and I fear these Losses may soon be productive of more and greater."[31] This last point was critical. Two weeks after Braddock's calamity, the commander at Fort Duquesne wrote to his superior, "I have succeeded in setting against the English all the tribes of this region who had been their most faithful allies. . . . They find themselves engaged in the war, so to speak, in spite of themselves."[32] Thrown on their own resources in the face of this war and without hope of an intercolonial defensive union, Pennsylvania and Virginia—the two principal colonies at risk from these attacks—turned more than ever to Franklin and Washington for military leadership.

BY 1755, Philadelphia was British North America's largest city and Pennsylvania its most prosperous colony for small family farms. These farms were spreading out far beyond the environs of Philadelphia to cover a broad arc of fertile land stretching from Easton and Bethlehem through Lebanon to York, with many of these communities populated by German-speaking immigrants, the so-called Pennsilfaanisch Deitsch or, later, Pennsylvania Dutch. With Braddock's army defeated and Dunbar showing no inclination to use its remaining rump for defensive purposes, these once secure central Pennsylvania communities and the scattered frontier settlements beyond them stood in the line of fire from France's Native allies.

Since its founding, Pennsylvania had relied on friendly relations with local tribes to expand through fair purchases and equitable treaties. More recently, several rapacious land deals (such as the infamous 1737 Walking Purchase, in which the proprietors' agent used runners and a prepared trail rather than forest walkers to

claim a vast tract under the odd terms of an old treaty that supposedly transferred land as far west as a man could walk in a day and
a half) had strained relations with the Natives. The French, now
imbedded with a line of forts within the colony's western border,
could inflame these tensions by supplying guns and ammunition
to the Natives.

Pennsylvania's peculiar politics added to the problem. The
colony's influential English Quakers retained their religious objections to warfare. They had lost absolute control over the colonial assembly, however, and most proved willing to pay taxes
for defense so long as others did the fighting. The frontier settlers
most at risk were unassimilated immigrants who spoke a foreign
tongue—prejudice against them ran deep. And the Penn family,
the colony's proprietors in London who appointed the resident
governor with veto power over laws, still refused to pay taxes on
its remaining land holdings even though, being mainly on the
frontier, they were at risk in any war with the French. In such a
crisis, with the frontier likely to explode after Braddock's defeat,
Pennsylvanians had learned to turn to one person for salvation.
Already larger than life, the proverbial self-made man, Franklin
was the inventor, orchestrator, producer, and promoter for any
public project. Although by no means an expert in military affairs
and notably down-to-earth for such an extraordinary person, he
had organized the colony's last-ditch defenses in the previous war
with France and stood ready to do so again. He became the man
of the hour.

Upon learning of Braddock's defeat, Pennsylvania governor
Robert Hunter Morris called the colony's assembly into session.
"There are Men enough in this Province to protect it against any
Force the French can bring," he declared, "but they have neither
Arms, Ammunition, nor Discipline, without which it will be im-
possible to repel an active Enemy whose Trade is War."[33] In short,

legislation was needed to raise and disburse funds for defense and to organize a militia. As simple as this might sound, all three steps split the colony along established fault lines. A quick way to raise revenue was by a property tax, but, since the proprietors refused to pay taxes on their holdings, the assembly (which held that all should pay their share) had not imposed one since 1717. When the assembly now promptly passed one for defense without exempting the proprietors' land, the governor, acting under orders from the proprietors, just as promptly vetoed it. The bill contained the added twist that the assembly would control disbursements, while the proprietors claimed this power for their appointed governor. A militia bill soon followed—the first in Pennsylvania history—but to meet objections from Quaker and antiproprietor members, it provided that participation was voluntary and that the enlisted men would elect (rather than the governor appoint) company officers and company officers would choose regimental commanders. These features further antagonized the proprietors.

The clear leader of the war party within the assembly and a principled foe of proprietary privilege, Benjamin Franklin stood behind these measures and the subsequent compromises that readied Pennsylvania for war. Indeed, during the legislative session, Franklin served on every committee dealing with the crisis, drafted the critical legislation, wrote the assembly's replies to the governor's messages, and put his spin on matters in the *Pennsylvania Gazette*.

"Why will the Governor make himself the hateful Instrument of reducing a free People to the subject state of Vassalage," Franklin wrote on behalf of the assembly in reply to Morris's veto of the bill taxing the proprietors' land for defense.[34] When the governor objected to the term "vassalage," Franklin added, "Vassals fight at their Lords Expense, but our Lord would have us defend his Estate at our own Expense! This is not merely Vassalage, it is worse."[35] In dozens of such exchanges over various war measures, Franklin got the better of Morris at every turn and foreshadowed the arguments later used against King and Parliament during the American Revolution. "Those who would give up essential Liberty, to purchase a little temporary Safety," Franklin wrote in defense of a revised revenue measure that retained the principle of equal taxation, "deserve neither Liberty nor Safety."[36]

Franklin and Morris dueled over the rights of English citizens versus principles of proprietary rule even as the warfare worsened. Cowed by the French and disenchanted with the British following Braddock's defeat, the Iroquois Confederacy largely remained neutral while the Delaware and other Natives allied with the French targeted isolated cabins and peripheral villages on the Pennsylvania frontier. First came word of a massacre at Penn's Creek—fourteen killed and scalped; eleven captured. The survivors petitioned the government about their land, "We are not able of ourselves to defend it for want of Guns and Ammunition."[37] Then a similar account from the Lebanon Valley, including the lurid detail of finding "four Indians sitting on Children scalping, 3 of the Children are dead, and 2 are alive with the Scalps taken off."[38] In November, the governor wrote from Philadelphia that Native war parties had pushed to within seventy miles of the city and "laid waste" to major farming settlements.[39] Refugees streamed in from the frontier and, in protest, at least twice placed bodies of scalped settlers on the State House steps. Then came news of a massacre of Moravian missionaries at Gnadenhütten, followed by the rout of a relief party sent to the site.

"We are all in Flames," Franklin wrote to Peter Collinson in England.[40]

Against the backdrop of defeat and despair, Franklin and Morris brokered the compromises that allowed for Pennsylvania's defense, particularly after the Quakers largely withdrew from the assembly. A contribution of income from past-due rents in lieu of a cash payment by the proprietors allowed a revenue measure to become law late in 1755, with Franklin serving as one of the commissioners charged with distributing the funds for defense. At virtually the same time, Morris consented to a voluntary militia with elected officers, with Franklin later chosen as the commanding colonel for the one-thousand-man Philadelphia regiment. Morris gave Franklin further power in January 1756 by placing him in charge of the large frontier county of Northampton, site of Gnadenhütten and other recent attacks, with total authority over its defenses—earning him the local title of "general."[41]

Franklin's fiftieth birthday, January 17, 1756, found him leading 150 soldiers north to Gnadenhütten, which they reached the next day. "Forty Dollars will be allow'd and paid by the Government for each Scalp of an Indian Enemy," Franklin told his men in line with a new official policy that outraged Quakers.[42] A larger bounty was placed on the heads of the Delaware chiefs. Taking extreme precautions against ambush, Franklin oversaw the construction at Gnadenhütten of a log stockade, 125 feet long and 50 feet wide, with two swivel guns and three interior cabins. When finished, he directed the erection of two more stockades, one fifteen miles east and one fifteen miles west, resulting in a line of forts garrisoned by some five hundred men securing Pennsylvania's northern frontier. By spring, settlers were returning to their farms in the region. Confronted with a military crisis unprecedented in Pennsylvania history, Franklin's strategy worked.[43]

No sooner had Franklin gotten this project well under way than he was called back to Philadelphia for the next legislative session, where he again led the antiproprietor forces intent on a stout defense. On his way from Virginia to Boston to meet the interim commander of British forces in America, Massachusetts governor William Shirley, George Washington reached Philadelphia at roughly the same time as Franklin and remained for a week. The two apparently met often to discuss frontier defenses. Franklin likely also raised his objection to the British army's enlisting indentured servants and apprentices without compensation to their masters, in the hope that Washington would convey his concern to Shirley. It was during Washington's stay that officers of the Philadelphia City Regiment elected Franklin as their commanding officer. Both men were now colonels in their colony's military and leaders in frontier defense. They met and corresponded frequently during the balance of 1756.

WITH VIRGINIA in as much immediate danger from the French as Pennsylvania, Washington had resumed command of his colony's military forces in August 1755. He had little choice. In response to the loss of Braddock's army, Governor Dinwiddie reconstituted the Virginia Regiment with an authorized force of a thousand men funded by the Virginia assembly. Nearly everyone looked to Washington as its leader.

Amid the flood of bad news for the frontier, his reputation for courage in battle had only grown, especially in the colonial press, where tales of his home-grown heroism contrasted with denunciations of British military ineptitude and cowardice. "Yor Name is more talked off in Pensylvenia then any Other person of the Army and every body Seems willing to Venture under your command," the frontier guide Christopher Gist wrote to Washington, adding that "Mr. Franklin" told him that if Washington asked the Pennsylvania assembly for aid defending the frontier, "you would now get it Sooner then any one in Amerrica."[44]

To accept command, Washington demanded control, and got it. His commission read, "Colonel of the Virga Regimt & Commander in Chief of all the Forces now rais'd & to be rais'd for the Defence of" Virginia, and it carried authority to appoint officers, buy supplies, and recruit soldiers.[45] Washington even designed the uniform for officers, which (in his words) "is to be of fine Broad Cloth: The Coat Blue, faced and cuffed with Scarlet, and Trimmed with Silver."[46] From London, he ordered ruffles and silk stockings for his own costume and livery suits emblazoned with his crest for his enslaved personal attendants. In the Virginia Regiment, Blacks could not serve as soldiers with guns but could die as servants without them. Officers proved easier to attract than soldiers, and soon the Virginia Regiment was top-heavy in command without ever securing a full complement of men, even after Virginia imposed a draft, which suffered the customary loopholes for those with wealth. Discipline proved a problem as well, leading Washington to lobby for, obtain, and sometimes impose the death penalty for desertion or disobedience. Until he secured that ultimate deterrent, he advised his chief lieutenant, Adam Stephen, "You must go on in the old way of whipping stoutly," and added the comment, "I must exhort you in the most earnest manner to strict Discipline, and due exercise of arms."[47] That was Washington's way: discipline and training.

During these months when each commanded forces in his own colony, Franklin exhibited a different style of leadership from Washington. Eschewing fancy dress and formal uniforms, Franklin spent his time on the frontier sleeping with his men on cabin floors and sharing food sent to the front by his wife. And when he heard that townspeople had prepared a hero's welcome for his return, Franklin slipped into Philadelphia after dark to avoid the show. Departing on a later date, he was surprised by a mounted escort that he did not request or want. Describing himself as "totally ignorant of military Ceremonies, and above all things averse to making Show and Parade," he attributed the display to his popularity—"the people happen to love me"—but went on humbly to allow that "Popular Favour is a most uncertain Thing."[48]

Washington evoked an opposite response, at least from many of those he ordered about. "In all things I meet with the greatest opposition no orders are obey'd but what a Party of Soldier's or my own drawn Sword Enforces," he said of local resistance to his martial-law orders, adding that once when commandeering a horse, its owner threatened "to blow out my brains."[49] Franklin, in contrast, gained cooperation by reasoned appeals and pragmatic solutions, such as when he boosted chapel attendance by authorizing the military chaplain to distribute rum after divine services. Where Washington's regiment was chronically undermanned, Franklin's was oversubscribed.

They had precisely the same job: to secure the frontier. For Washington as much as Franklin, this meant the unglamorous task of building and maintaining a line of forts on the wilderness edge of civilization, and trying to keep hostile forces out. In the defense of Virginia, Fort Cumberland in Maryland served as the northern terminus of this line, which stretched south through the Shenandoah Valley to the North Carolina border. Winchester, then Shenandoah's only town and site of one of his farms, served as headquarters for Washington. He dreamed of marching on Fort Duquesne and cutting off the problem at its head but never had enough men to defend the border fully from Native raiders, much less carry the war to the French.

Washington's initial series of meetings with Franklin in 1756 resulted from a row over rank at Fort Cumberland, pitting his Virginia colonelcy against a Maryland captain with a regular army commission, that Washington carried all the way to the supreme British commander in Boston. He met with Franklin on the journey both up and back, and later when Franklin traveled to Virginia on post office business. These meetings were cordial but rushed, as each had other places to go or business to accomplish, but they led to an exchange of letters, most of which are lost. In one, Franklin alerted Washington that "the Delaware Indians," as he called them, had rejected Pennsylvania's peace offering "and resolv'd to continue the War."[50] During their meetings, Franklin and Washington likely discussed strategy and tactics at length, with Washington coming to espouse a Franklinian view of inter-colonial military cooperation. For his part, Franklin always saw Washington as the better commander.

Virginia never suffered any single massacre as large as some in Pennsylvania but, with its settlers scattered out over a wider area in more isolated farmsteads, lost more total people killed or captured. "They go about and Commit their Outrages at all hours of the day and nothing is to be seen or heard of, but Desolation and murders," Adam Stephen reported to Washington about the Natives around Fort Cumberland in October 1755. "The Smouk of the Burning Plantations darken the day, and hide the neighbour-ing mountains from our Sight."[51]

One classic ambush, in April 1756, cost the lives of two officers and fifteen soldiers—"a very unlucky affair," Washington called it[52]—amid a springtime surge in raids that terrorized the pop-ulation. Two years later, a similar number of Virginia soldiers died when two contingents of them, one led by Washington, mistakenly fired on each other in the wilderness and continued doing so until Washington, as he remembered it, rushed "be-tween [the] two fires, knocking up with his sword the presented pieces." With bullets again whistling around him, he felt the pro-tective hand of providence. At that moment, Washington long after wrote, his life was "in as much jeopardy as it had ever been before or since."[53]

Unlike Pennsylvania, Virginia had the added worry that raiders would cross the mountains to incite slave revolts in long-settled regions, which forced the colony to keep local militia units on guard in their home counties. Virginia also had a history of worse relations with the Native people than Pennsylvania and lacked the mediating influence of Quakers, who struggled to maintain a dialogue with the Delaware throughout the conflagration. Treaties quieted the Pennsylvania frontier in 1756 even as the situation worsened in Virginia. Ever the realist, Franklin was on record doubting whether Pennsylvania would "ever have a firm Peace with the Indians until we have drubb'd them,"[54] but fully supported a negotiated solution and had earlier written of "the lovely White and Red" increasing together in America,[55] while Washington always viewed removal as the only solution to the so-called Indian problem.

NEITHER FRANKLIN NOR WASHINGTON saw the war through to its victorious conclusion in 1763, when all of Canada and the Ohio Country passed from French to British rule along with Florida going from Spain to Britain and Louisiana from France to Spain, but each remained involved until stability returned to his colony's frontier. Franklin left first, following his role in negotiating the Easton Treaty with the Delaware, when Pennsylvania's assembly chose him in 1757 as its agent to press its grievances against the proprietors in London and perhaps overthrow their rule altogether. Already a renowned scientist, Franklin regarded this extended sojourn, which covered fifteen of the next seventeen years, as a chance to circulate on an international stage. His final wartime conference with Washington and other colonial military leaders came in February 1757, shortly before he sailed for England. Franklin had planned to depart earlier but the supreme British commander requested his attendance. Washington used the occasion of this conference to renew his petition for a regular British officer's commission, adding the telling plea on behalf of himself and his men, "We cant conceive, that being Americans shoud deprive us of the benefits of British Subjects; nor lessen our claim to preferment."[56] By this time, he very much did view his American status as a black mark in British eyes, and resented it.

In April 1758, Washington returned west for one last season at the front. This stint culminated in autumn with a march on Fort Duquesne by a seven-thousand-man force composed mostly of troops from five middle colonies. By this time, the colonial French and Indian War, ignited in part by Washington's rash attack at Jumonville Glen, had exploded into the global Seven Years' War involving all the great powers of Europe and their colonies on five continents. Britain pressed French Canada on multiple fronts. Those assaults coupled with the Easton Treaty left French control of Fort Duquesne unsustainable. The French abandoned it before the British arrived in November, and with it their ability to support raids into Virginia.

In the meantime, Washington had come into his own as a fullfledged member of the Virginia landed gentry. During 1758, his final year of colonial military service, he expanded Mount Vernon; became engaged to a wealthy widow, Martha Dandridge Custis, with vast dower holdings in land and slaves; and was elected to Virginia's assembly, the House of Burgesses. With the frontier secure and his domestic prospects bright, he resigned his military post for the life of a country squire.

Each then having defined lives in distinct positions on distant shores, the budding relationship between Franklin and Washington went into hibernation. At the time, of course, Franklin was still twice Washington's age and, with the war over for them, they had no need to meet or correspond. Once another war loomed, however, that would change.

They had both learned similar lessons from the French and Indian War. First, the British had different objectives from their American colonists. The British wanted to keep the colonies divided and dependent, Franklin concluded from his experience with the Albany Plan of Union, and would gladly tax them without representation. Washington found the British unwilling to secure the frontier except as it served their larger geopolitical interests. Second, American colonists would always remain subordinate to their British counterparts. Washington's bitter frustration with rank confirmed this. He pleaded for a royal commission without receiving one and was compelled to submit to inferior British officers. As for Franklin, when Pennsylvania's proprietors (who still ran the colony within broad parameters set by Parliament) heard of his appointment as the assembly's agent in London, they dismissed his possible influence in England. "Mr. Franklin's popularity is nothing here," Thomas Penn coolly commented. "He will be looked very coldly upon by great People."[57] Third, the American colonies would benefit from greater unity as reflected in Franklin's visionary Albany Plan and Washington's calls for joint intercolonial military action. After all, it was a force composed mainly of troops from five colonies that finally drove the French from Fort Duquesne. This experience made Washington, as much as Franklin, a believer in union.[58]

These three lessons might suggest benefits from American independence but were insufficient to support it as a realistic option. A final shared lesson carried more weight. Despite the war's ultimate outcome, the British were beatable in New World combat. "This whole Transaction gave us Americans the first Suspicion that our exalted Ideas of the Prowess of British Regulars was not well founded," Franklin wrote of Braddock's defeat.[59] Washington had been there to see it and to report that, at least in frontier fighting, Virginia soldiers outperformed British troops. If put to the test, they might do so again. Coupled with the disastrous effects of British colonial policy following the French and Indian War, these shared lessons helped to nurture the revolutionary spirit that would bring Franklin and Washington back together a quarter century later to fight for and forge a new American nation.

Three

FROM SUBJECTS TO CITIZENS

FOR HALF A DECADE after the end of their participation in the French and Indian War—even as that increasingly global conflict dragged on until 1763—Franklin and Washington took different paths. Seeking to overthrow proprietary rule in Pennsylvania and replace it with a royal charter akin to those of most other British colonies, Franklin served as the agent for his colony's assembly in London. He returned to Philadelphia for two years, 1762–1764, but soon was back in London as the agent not only for his colonial assembly but for others too. His fame and wit won him entrance to elite salons, the ear of government ministers, and friendship with philosophers and scientists, but his best efforts could not dislodge the Penn dynasty from its proprietorship. Meanwhile, faced with mounting debts due to depressed tobacco prices, Washington struggled to maintain his planter lifestyle by carefully managing his expanding plantation, diversifying its products, and speculating in frontier lands. That and a rising role in Virginia's House of Burgesses kept him close to home.

Although both men gloried in their Englishness and rejoiced in the victory that brought all of eastern North America from Florida to Canada under British rule, their paths might have never crossed again save for the Stamp Act crisis, which erupted in 1765 after Parliament imposed taxes directly on the colonists. It jarred their interests back into alignment, set them on parallel political courses, and reminded them of their essential Americanness. Yet neither saw it coming.

By the fall of 1765, riots tore through many of America's major towns, touched off not by foreign invaders but by homegrown tax resisters calling themselves Sons of Liberty.[1] The liberty they claimed was the supposedly age-old right of the English to be taxed only by their own representatives, not by a parliament with none of their elected members. Both Franklin and Washington had espoused this right in the years leading up to the Stamp Act crisis, and Franklin had argued for colonial representation in the British Parliament as a cure, but neither of them appreciated the depth of the right's popular appeal. Strangely tone-deaf prior to the crisis and nearly left behind by the popular surge, both managed to get themselves ahead of the crowd and assume a moderating leadership role in the end. In the process, they began gradually shifting their loyalty from imperial Britain to republican America.

THE CRISIS RESTED on a fundamental difference: Parliament viewed colonists as its subjects; colonists viewed themselves as British citizens. On this score, Franklin and Washington held the colonists' view. This difference might have remained academic had a new British government not sought to balance the postwar budget on the back of colonists. True, the war had cost Britain dearly, but it had also cost the colonies, in both blood and treasure. True, British armies had fought in America to conquer Canada, but so had colonial forces. True, the British victory had removed the French threat from the colonies' frontier, but the colonists saw themselves as partners in that victory and Britain as gaining the most, particularly after the Proclamation of 1763 barred colonial settlement west of the Appalachian Mountains and the 1764 Quebec Act appended the Ohio Country to Canada. True, colonists did not pay taxes directly to Britain other than the regulatory import tariffs common to all within the Empire, but they paid taxes to their colonies, which in turn contributed to imperial causes. While people might disagree on the justice of taxing unrepresented colonists, self-interest inclined colonists to see it as unjust. After the fall from power of William Pitt, whose imperial strategy relied on cooperating with the colonies, and the rise of George Grenville as prime minister, those in favor of taxing colonists gained control of Parliament.

Ruling a vastly expanded British Empire after the Treaty of Paris ended the Seven Years' War in 1763, Grenville proposed a comprehensive scheme of imperial administration that a critical mass of American colonists deplored. Where many earlier ministries adopted a sort of benign neglect toward the colonies by valuing them mainly as captive markets for trade, Grenville regarded them as subservient parts of an empire in desperate need of revenue. The war had left the British government with an enormous debt. The peace saddled it with the ongoing expense of militarily occupying such restive new domains as French Canada, Spanish Florida, and a Great Lakes region that promptly erupted in a widespread uprising of Native peoples initiated by the Ottawa leader Pontiac. That clash left nearly five hundred British soldiers, two thousand settlers, and countless Natives dead. Adding to these costs, Grenville sought to ensure submissive colonial governors and officers by having London pay their salaries directly rather than leaving it to their colonial assemblies. Under his scheme, colonists would foot the bill for their own administration and defense through new revenue tariffs on goods imported into the colonies and tax stamps on paper used for court filings, commercial transactions, and a wide range of printed items, from newspapers to playing cards. Even pamphlets protesting the Stamp Act would be taxed. An additional act authorized the quartering of British troops in colonial homes and buildings.

Although he unveiled his full plan in 1764, fearing focused opposition to it as a direct, internal excise, Grenville delayed a parliamentary vote on the stamp-tax provision until the next year to allow the colonies and their agents time to offer alternatives. As Grenville hoped, the new revenue tariffs generated little initial opposition in the colonies, presumably because they seemed like the old regulatory tariffs—the main change being that British agents now tried harder to collect them. Known in America as the Sugar Act because of its tariff on foreign sugar, the American Duties Act of 1764 elicited official protests only from New York and Massachusetts. James Otis and Samuel Adams stirred the pot in the latter, but the formal petition from Boston did not even raise the thorny issue of taxation without representation. Otis and Adams had harped on it of course, but Thomas Hutchinson, the colony's loyalist lieutenant governor, turned the petition into one about the act's harsh impact on trade rather than its constitutionality.

Trade was an issue too, providing an essential context for colonial resistance to British taxation, because peace with France had not brought prosperity to the colonies. Quite to the contrary, the end of wartime military spending coupled with a European bank panic triggered a business downturn that began in northern ports during the early 1760s and gradually engulfed all the colonies in an economic depression that persisted for most of the decade. Britain's decision with the Currency Act of 1764 to protect its merchants by barring the colonies from issuing new paper money and directing them to phase out old issues made matters worse by restricting the money supply that fueled the local economy. Virginia, which at the time had nearly a quarter million pounds of such notes in circulation, was especially hard hit as its money supply shrank. Sinking in debt to his London business agent and without money to repay it, in 1764 Washington began the process of turning his plantation from raising tobacco for export to growing wheat for domestic sale. America could get along with less reliance on Britain, he started to grasp. In 1763, Washington had hailed news of peace with France by stating his hope "that the Tobacco trade will fall into an easy and regular Channel again." But he soon realized that it would not happen under emerging British trade, tax, and monetary policies.[2]

THE EVOLUTION IN COLONIAL ATTITUDES toward Britain that began with the Sugar Act, Currency Act, and Quebec Act in 1764 turned toward revolution with passage of the Stamp Act in 1765. No one could disguise it as anything but a direct, internal tax on American colonists. Grenville did not seek cover. He wanted to establish the principle of parliamentary sovereignty over the colonies. Washington and Franklin dreaded this. In December 1764, during the comment period after Grenville proposed the stamp tax but before Parliament voted on it, Virginia's colonial assembly, with Washington present, raised the volatile issue in a petition to King George III. It asked him to protect the "People of this Colony in the Enjoyment of their ancient and inestimable Right of being governed by such Laws respecting their internal Polity and Taxation as are derived from their own Consent."[3] Addressing longer pleas to the House of Lords and House of Commons, the House of Burgesses declared "it to be a fundamental Principle of the British Constitution, without which Freedom can no Where exist, that the People are not subject to any Taxes but such as are laid on them by their own Consent, or by those who are legally appointed to represent them."[4] Washington expressed support for these petitions in letters to family and friends.

Pennsylvania's assembly also formally objected to the proposed

Stamp Act and, in November 1764, dispatched Franklin back to London as its agent to oppose it. He had been home for two years, during which time he resumed active administration of the colonial postal service. This included adding routes to Canada and overnight delivery between Philadelphia and New York. He also began constructing a new house, the first that he owned rather that rented; started speculating in western lands; and saw his son William installed as the royal governor of New Jersey. Mainly, however, he reclaimed his seat in the assembly, became its speaker, and then lost his seat by nineteen votes in the complex battle over proprietary rule. His faction retained power, however, and promptly returned him to his post in London, where he now had to balance seeking a royal charter for Pennsylvania with slamming parliamentary measures infringing on the rights and interests of colonists. This proved a tricky minuet because, to some, the old sins of Penn's descendants paled in contrast with the new ones of Parliament and a royal charter would put Pennsylvania more directly under parliamentary rule.

Shortly after Franklin arrived in London, Grenville called him and other colonial agents together to discuss the proposed stamp tax. "Mr. Grenville gave us a full hearing," one of them reported to his governor, but maintained that the colonies "ought to pay something, & that he knew of no better way than that now pursing to lay such Tax, but that if we could tell of a better he would adopt it."[5] Franklin could and did but, "being besotted with his Stamp Scheme" (as Franklin put it), "Grenville paid little Attention to it."[6]

Displaying his usual cunning, Franklin had proposed that, instead of a general stamp tax, Britain levy a tax (in the form of interest) on paper money printed by the colonies. This would at once shift the burden onto voluntary acts of colonial governments, thus avoiding a clear assault on individual rights, and address the economically devastating effects of tight money by effectively reversing the Currency Act. In part because he wished to establish the power of Parliament to tax colonists, Grenville ignored Franklin's compromise and pushed the Stamp Act through the House of Commons on February 27, 1765. It became law a month later.

Only one member of the Commons rose in strong defense of colonists' rights. His speech to Parliament became a rallying cry in America. Warning that the Stamp Act would create "disgust, I had almost said hatred," Isaac Barré spoke warmly of "those Sons of Liberty" who "Actuated by principles of true english Lyberty" had carved British colonies out of "a then uncultivated and unhospitable Country." Barré concluded his address with the prophetic observation that, from his experience with them as a British officer during the Seven Years' War, the colonists in America "are as truly Loyal as any Subjects the King has, but a people Jealous of their Lyberty and who will vindicate them, if ever they should be violated."[7] Reprinted widely throughout the colonies, Barré's artful phrase "Sons of Liberty" was embraced by those colonists who rose in opposition to the Stamp Act and subsequent efforts to tax the colonies.

ALTHOUGH OPPOSED TO THE STAMP ACT from the outset, Franklin and Washington failed at first to foresee its significance. As the historian Edmund Morgan has noted, Franklin cared more about doing what was right than defending abstract rights. To him, the Stamp Act's predictable adverse impact on commerce and imperial relations made it wrong for Britain and the colonies. Being in London, however, Franklin could not feel the resistance that the rights-based argument was stirring in the hearts and minds of colonists from Boston to Charleston. Not sensing this, once the act passed, he tried to salvage what he could from the new law by using his position as assembly agent to secure the lucrative post of stamp distributor for Pennsylvania and Delaware for the assembly's speaker, John Hughes. Better that the job go to a political ally than to a lackey of the proprietors, Franklin reasoned. By appearing complicit in the administration of the Stamp Act, however, he made one of the worst blunders of his career. Although less directly implicated than Franklin, Washington also suffered some taint when George Mercer, an old friend, former military aide, and current business partner, was named the stamp distributor for Virginia. Once the issue became a fight for principle, Franklin and Washington found that anything less than purity was suspect.

The troubles began in Virginia where, after word reached America that Parliament had passed the Stamp Act, Patrick Henry, a self-made, silver-tongued lawyer newly elected to the House of Burgesses, persuaded that assembly to approve five new resolutions against it. By this point, at Otis's urging, Massachusetts had called for an intercolonial congress to consider a unified response to the issue. A spellbinding orator with radical views, Henry now gave it a hard edge with Virginia's resolves. The first four largely restated positions that the assembly had staked out in its earlier pleas. The fifth inched further by declaring that the colony's assembly had the "Sole Right and Authority to lay Taxes and Impositions upon It's Inhabitants: And, That every Attempt to vest such Power in any other Person or Persons whatsoever, has a Manifest Tendency to Destroy AMERICAN FREEDOM." By various accounts, none of them authoritative, Henry secured passage of the resolves with an impassioned speech that bordered on sedition ("Caesar had his Brutus—Charles the First, His Cromwell—And George the Third . . ." was how legend remembered it) before House of Burgesses speaker Peyton Randolph and others shouted him down with cries of "Treason!" To this, Henry reportedly replied, "If this be treason, make the most of it."[8]

Henry apparently had two more resolutions in reserve but, after the fifth barely passed, he did not offer them. All seven resolutions appeared in some published lists of those passed. As printed in many newspapers, a sixth resolve declared that colonists "are not bound to yield Obedience to" an improper tax. The seventh held that any person enforcing such a tax "shall be Deemed, AN ENEMY TO THIS HIS MAJESTY'S COLONY."[9] Washington likely went home before these end-of-session resolutions passed, and after Henry left the assembly expunged the fifth one. By then, however, Virginia had set the tone for stiff resistance to the Stamp Act. All seven resolves were widely discussed, and Rhode Island's assembly passed ones modeled on them, including the sixth but not the seventh.

Boston exploded first. Organized working-class mobs invoked by a middle- and upper-class group of artisans, merchants, and lawyers, the so-called Sons of Liberty, destroyed the homes of Hutchinson and Andrew Oliver, the colony's designated stamp distributor. Some protesters threatened worse if Oliver did not resign. He did. Taking refuge in an offshore British fort, the colony's royal governor ordered drummers to call out the local militia to restore order, only to learn that the drummers had joined the mob. So had much of the militia. The sheriff refused to intervene. All attempts to enforce the tax proved futile. Calling in British troops would likely make matters worse, the governor feared. Commerce soon resumed and courts reopened as if the act had never passed.

Similar scenarios played out everywhere. In Franklin's Philadelphia, Hughes brandished arms to deter a mob threatening his house but ultimately agreed not to enforce the Stamp Act in Pennsylvania unless other colonies implemented it. They didn't. Egged on by a pro-proprietor Presbyterian faction within the fragmented and fractious world of Pennsylvania politics, the mob even threatened Franklin's house, which his wife vowed to defend in his absence. It was her home too, of course.

"I sente to aske my Brother to Cume and bring his gun," she reported to her husband in London. "So we maid one room into a Magazin. I ordored sum sorte of defens up Stairs such as I Cold manaig my self."[10]

Signing himself "your ever loving husband," Franklin wrote back, "I honour much the Spirit and Courage you show'd, and the prudent Preparations you made." Blaming the leader of the Phila- delphia Presbyterians for circulating a rumor that he planned the Stamp Act, Franklin sarcastically added, "I thank him he does not charge me (as they do their God) with having plann'd Adam's Fall, and the Damnation of Mankind. It might be affirm'd with equal Truth and Modesty."[11]

Shunting aside Hughes as its leader, Franklin's faction in the assembly promptly declared the Stamp Act unconstitutional in Pennsylvania and blamed the entire mess on the proprietors and their party.

The confrontation took a different form in Washington's Vir- ginia because Mercer, the stamp distributor, was in London when the Stamp Act passed and at sea with the stamps as riots broke out elsewhere. He was greeted at the dock in Williamsburg by two thousand Virginians demanding his resignation. After consulting with the royal governor, who suggested that he step down if he feared for his life, Mercer accepted the inevitable. No stamp taxes were ever collected in Virginia or any of the original thirteen mainland colonies. The assemblies of nine colonies passed reso- lutions claiming the exclusive right to tax their citizens—others being prevented from doing so by their royal governors—and nine sent delegates to the Stamp Act Congress in New York, which adopted similar resolves. In both a budding reaction to the Sugar Act and a strategic response to the Stamp Act, boycotts of British products spread.

Washington lamented Mercer's rough treatment by the mob but sent an angry letter to his business agent in London denouncing imperial policy in terms that echoed the mob. "The Stamp Act," he wrote, "engrosses the conversation of the speculative part of the Colonists, who look upon this unconstitutional method of Taxation as a direful attack upon their Liberties." Knowing that his agent had influence with Parliament and hoping to turn him against the act, Washington explained that, so long as this "ill Judgd measure" is in force, the courts (which required stamped papers to act) will close. He called it "morally impossible" for Virginia lawyers and judges to comply with it. With courts needed to collect debts, Washington then pointedly asked, "Who is to suffer most in this event—the [British] Merchant, or the [Virginia] Planter?" Expanding his critique to include the Sugar Act, Washington embraced the growing nonimportation movement. "The Eyes of our People," he wrote, are "already beginning to open" that domestic products can replace British ones. Washington closed with a question that answered itself: "Great Britain may then load her Exports with as Heavy Taxes as She pleases but where will the consumption be?"[12]

FOR REASONS WHOLLY UNRELATED to the Stamp Act fiasco, the more moderate Marquis of Rockingham, a Pitt ally, replaced Grenville as British prime minister in July 1765. With full support from London merchants, Rockingham wanted nothing more than to repeal the divisive and ineffective Stamp Act, but he needed to proceed cautiously in light of the still powerful Grenville faction, which refused to concede authority over taxes to the colonies.

For help, Rockingham turned to Franklin, who took center stage in Parliament's artfully choreographed retreat. Franklin was the best-known and most respected colonist in London, and recognized as a moderate on imperial issues—so much so that anti–Stamp Act mobs had threatened his house. He was, in short, the ideal witness to testify for repeal. With Parliament sitting as a committee of the whole, Rockingham invited him to stand in the well of the House of Commons as an expert on American public opinion. The result was a remarkable three-hour-long performance during which Franklin fielded friendly questions from Rockingham, William Pitt, and backers of repeal along with furious queries from Grenville and his allies.

Franklin painted a telling portrait. Loyal but aggrieved, colonists were ready to submit to Parliament, he said, on virtually all matters except internal taxes. "They had not only a respect, but an affection, for Great-Britain, for its laws, its customs and manners, and even a fondness for its fashions, that greatly increased the commerce," Franklin said of the colonists' temper before 1763. "They were governed by this country at the expence only of a little pen, ink and paper. They were led by a thread." But after the Stamp Act, their temper was "very much altered."[13]

Nothing will get them to pay this tax, he declared, not even armed soldiers sent for that purpose. "They cannot force a man to take stamps who chooses to do without them. They will not find a rebellion; they may indeed make one," he warned.[14] Having long argued that mob violence would only serve to stiffen British resolve, Franklin stressed the peaceful nature of the protests. The assemblies themselves would punish the rioters "if they could," he said.[15] Rather than threaten that retaining the Stamp Act *might* lead to open revolt, he averred that it *would* result in expanding boycotts of British goods. This was what proponents of its repeal wanted to hear.

Focused on securing that repeal, Franklin drew an expedient but unworkable line between unacceptable internal excises and acceptable external duties. When asked about acceptable taxes by friendly questioners, he spoke with care of duties designed to regulate trade, but when pressed by Grenville or other hostile interrogators, he expanded them to include revenue duties as well. "The sea is yours," he said at one point, "you may have therefore a natural and equitable right to some toll or duty on merchandizes carried through that part of your dominions."[16] Franklin knew that in principle the Sugar Act was as objectionable as the Stamp Act to many colonists and that some of the petitions to Parliament condemned both, but when faced with such evidence he dismissed it. "They mean only internal taxes," Franklin said of petitions that rejected all taxes imposed by Parliament, "the same words have not always the same meaning here and in the Colonies."[17]

An eminently practical politician, Franklin accepted the fateful political compromise that would expunge the Stamp Act in exchange for retaining (and later expanding) external revenue tariffs—a compromise that kept the revolutionary spirit alive in America without placating imperialists in Britain. In a rehearsed response, Franklin also affirmed that, so long as Parliament did not actually impose any internal taxes on them, colonists would give "very little concern" to a face-saving gesture, the so-called Declaratory Act, asserting that Britain had the right to do so.[18] Without any evidence on this point, he simply asserted that colonists were as pragmatic about their principles as he was about his.

When Parliament repealed the Stamp Act in 1767, colonists from Massachusetts to the Carolinas praised Franklin for his role in the process without realizing the concessions that he offered in the process. Focusing on the repeal of the Stamp Act rather than the continuation of the Sugar Act or passage of the Declaratory Act, Washington hailed the result as good for Britain and America. "All therefore who were Instrumental in procuring the Repeal," he wrote with Franklin presumably in mind, "are entitled to the Thanks of every British Subject & have mine cordially."[19]

Historians generally see the Stamp Act crisis as a pivotal moment in American popular politics and political thought. For the first time, the colonists, realizing that their interests were not represented in London, began thinking of themselves more as citizens of their colonies under the king than as subjects of Parliament. Franklin and Washington observed this important shift in others and felt it within themselves. While they remained loyal to the crown and hoped that the Empire would endure, both grew wary of parliament.

Privately, Franklin finally gave up on his idea of representation for the colonies in Parliament, seeing all parties as indifferent or hostile to the notion. When speaking about the colonists in his testimony before the House of Commons, he duly recited that "they consider themselves as a part of the British empire, and as having one common interest with it," but added the critical caveat that "they may be looked on here as foreigners."[20] As Poor Richard might say, alienation breeds estrangement.[21]

The uniform American response to the Stamp Act united the colonies. "Such a Union was never before known in America," John Adams noted at the time. "In the Wars that have been with the French and Indians a Union could never be effected." Foreseeing its value, both Franklin and Washington had called for intercolonial cooperation during the French and Indian War. Now its value was widely apparent. For thirteen small colonies, power lay in unity, not division. Georgia, New Jersey, and Massachusetts asked Franklin to serve as their agent in London, along with his work for Pennsylvania, and both Washington and Franklin began speculating in frontier land beyond the boundaries of their own colonies. By working together to oppose the Stamp Act, the colonies had prevailed. Thus, while their boycotts wound down once Parliament repealed the act, they never completely ended. Concerns remained about the Declaratory and Sugar Acts, with some like George Mason, Washington's neighbor and a former member of the House of Burgesses, seeing them as part of the same grand British plot against individual liberty that produced the Stamp Act. Washington began thinking that way too, and it was reflected in the resolution passed by the Virginia assembly after the Stamp Act's repeal, which still spoke in terms of defending "just Rights and Liberties" and opposing "unconstitutional" violations.[22]

Beyond discovering the surprising resolve of colonists from all stations of life to stand up for their rights, Franklin and Washington saw in the Stamp Act crisis a means for the colonies to assert their interests peacefully through boycotts. "Many of the Luxuries which we have heretofore lavished our Substance to Great Britain for can well be dispensed with whilst the Necessaries of Life are to be procurd (for the most part) within ourselves," Washington wrote during the height of the crisis. "This consequently will introduce frugality; and be a necessary stimulation to Industry."[23]

Franklin made a similar point in his testimony before Parliament. "The goods they take from Britain are either necessaries, mere conveniences, or superfluities," he said of the colonists. "The first, as cloth, &c. with a little industry they can make at home; the second they can do without." As for the third, he added perhaps with an impish grin, "They are mere articles of fashion, purchased and consumed because [they are] the fashion in a respected country, but will now be detested and rejected."[24] The beauty of a boycott, both men agreed, lay in its making colonists more economically independent even as it upheld their political rights. Over the next decade, nonimportation associations became favored tools of Franklin and Washington for combating imperial oppression.

IN 1767, less than a year after the Stamp Act's repeal, Washington planted his first full crop of wheat at Mount Vernon. To stay afloat financially, he transformed his seven-thousand-acre estate from a tobacco plantation into a grain farm. British policies and practices made it hard for large plantations like Mount Vernon, with marginal soil and hundreds of slaves, to turn a profit on tobacco. With the overseas tobacco trade required by law to pass through England, much of the revenue went to London agents. Many planters sunk deep in debt. Washington turned to domestic products, such as wheat, corn, and livestock. He built a gristmill to grind his grain and, for a fee, his neighbors' crops as well. With fewer slaves needed to grow grain than raise tobacco, he redeployed some of them to such profitable enterprises as weaving, fishing, and distilling. In this way, Mount Vernon became virtually self-sufficient, other than in luxury goods. Heavily indebted to his London agent at the onset of the postwar depression in 1764, by embracing the Franklinian virtues of industry and frugality (though in practice more of the former than of the latter), Washington cut his unpaid obligations in half by 1770.

The shift from export tobacco to domestic grains made it easier for Washington to boycott British goods when the need arose again. Further, as he noted in a letter to George Mason, nonimportation furnished "a pretext to live within bounds," where otherwise "such an alteration in the System of my living, will create suspicions of a decay in my fortune, & such a thought the world must not harbour."[25]

Colonists such as Franklin may have believed, and those such as Washington hoped, that repeal of the Stamp Act would signal the end of imperial experiments with colonial taxes, but some feared otherwise, the conspiracy-minded Mason included. These doubters had reasons to fear the worst. The Declaratory Act asserted Parliament's sovereignty over the colonies in all cases whatsoever. If the supposed bright line between unacceptable internal and acceptable external taxes held, then Parliament could (and most likely would) impose whatever duties it wished on colonial trade. It only took the fall of Rockingham and the rise of the Grenville-like Charles Townshend as chancellor of the exchequer in 1767 for Parliament to pass the so-called Townshend Act imposing revenue duties on tea, paper, glass, lead, and paint entering the colonies. Compounding the problem, as a Tory backer of the ambitious young Hanoverian king George III, Townshend was viewed by some radical Whigs as a witting or unwitting pawn in a plot to undermine traditional English liberties and impose continental absolutism at home as well as in the colonies. Taxing colonists without representation, these Whigs feared, was the first step in a dark conspiracy. Libertarian-minded colonists like Mason drank in these fears and shared them widely.

Listening to Mason and perhaps believing him, for the first time Washington, a cautious Anglophile with nerves of steel in combat, took a public stand against British imperial policy. Starting in early April 1769, he began working with Mason to forge a nonimportation association for Virginia, similar to those in other colonies, to boycott nonessential British goods so long as Parliament persisted in taxing the colonies. In 1768, the Virginia assembly had sent petitions to the king and Parliament protesting the Townshend Act, but received no relief. Now Washington wanted to ratchet up the pressure with a boycott and voiced his willingness to resort to force if necessary.

"At a time when our lordly Masters in Great Britain will be satisfied with nothing less than the deprivation of American freedom, it seems highly necessary that something shou'd be done to avert the stroke and maintain the liberty which we have derived from our Ancestors," Washington wrote to Mason in words that he surely knew would resonate with Mason's conspiratorial thinking. He would not "hesitate a moment to use a[r]ms in defence of so valuable a blessing," Washington declared with a flourish, but this "should be a last resource." Petitions had proved worthless with the British. Now was the time to draw "their attention to our rights & priviledges . . . by starving their Trade & manufactures" with a boycott, Washington wrote.[26]

Washington enclosed a copy of a nonimportation agreement adopted by Philadelphia merchants and asked for Mason's help in adapting it for use in Virginia. In it, Pennsylvania merchants vowed to boycott all British imports—not just taxed items—except for twenty-two listed essential commodities. Boston and New York merchants had already formed similar nonimportation associations. Mason readily agreed. "Our All is at Stake, & the little Conveniencys & Comforts of Life, when set in Competition with our Liberty, ought to be rejected not with reluctance but with Pleasure," he replied to Washington.[27] The two men began working together to have a nonimportation resolution ready for Washington to present at the next session of the House of Burgesses, in May 1769.

Washington lacked Franklin's subtlety—that is, the Virginia planter was more fixed in his ends and straightforward in his means than the Sage of Philadelphia. Once Washington declared his course, he pursued it doggedly and labored to bring others along. Committed now, he carried to Williamsburg the text of the resolution forming a nonimportation association for Virginia.

Their proposal turned the Philadelphia agreement on its head. Rather than suspend the importation of all goods with a few stated exceptions, it listed the boycotted items, which its drafters selected to impose maximum pain on Britain with minimum injury to Virginia. Mason wanted to add the threat that, if the Townshend Act remained in effect, Virginians would stop exporting tobacco—but Washington apparently found that idea impractical. Like the Philadelphia agreement, the boycott would last until repeal of all of the Townshend duties. In Philadelphia, Boston, and New York, where trade passed through cities, only merchants and traders joined the nonimportation associations. In Virginia, with its large plantations and network of navigable tidewater rivers, consumers like Washington typically ordered imports directly from London for delivery to their riverfront docks. This difference led Washington and Mason to address their agreement to members of the Virginia assembly, which included many of the colony's richest landowners and principal importers, with the goal of having them sign it and carry copies to their home counties for others to sign.

Seeking broader participation than those for northern port cities, Virginia's nonimportation association was expressly designed for "all Gentlemen, Merchants, Traders, and other Inhabitants of this Colony" to join.[28] Signers agreed to instruct their British correspondents not to supply them with goods either subject to the Townshend duties or on the enumerated list of banned imports. That was the political pressure point: British merchants and manufacturers. Doing his part, Washington referenced the agreement in his next order for goods from his London business agent, adding the instruction, "I am therefore particular in mentioning this matter as I am fully determined to adhere religiously to it, and may perhaps have wrote for some things unwittingly which may be under these Circumstances."[29] Do not send him any boycotted items, he told his agent.

Washington arrived in Williamsburg for the assembly's May session at a pivotal moment in Anglo-American relations. Digging in its heels on the principle that it alone could tax Virginians, the House of Burgesses passed another resolution on the morning of May 16 reasserting its rights. Dispatched to Virginia to defend the principle of Parliament's sovereignty in all matters, the colony's new royal governor, Baron de Botetourt, by noon on the same day disbanded the assembly.

"I have heard your Resolves, and augur ill of their Effect," Botetourt admonished the astonished members in words that they never forgot. "You have made it my Duty to dissolve you; and you are dissolved accordingly."[30]

Rather than quietly disperse, virtually all members of the Virginia assembly promptly reconvened at nearby Raleigh Tavern. After discussing matters, they appointed a committee that included Washington to prepare a nonimportation association agreement for ratification on the following day. Apparently Washington had been shopping around the plan that he had drafted with Mason. Probably still in shock from the governor's abrupt action, 87 of the assembly's 116 members signed the nonimportation agreement at that next meeting and agreed to circulate copies for added signatures.

Having taken their stand, they drank toasts both to "A speedy and lasting Union between *Great Britain* and her colonies" and to "The constitutional *British* Liberty in America, and all true Patriots, the supporters thereof."[31] Still favoring union with Britain, these men now saw themselves as patriots to a higher cause. With Washington assuming a lead role for the first time since the French and Indian War, they vowed to defend liberty in America through joint action with other colonies. Before leaving Williamsburg, Washington bought a new book, *Letters from a Farmer in Pennsylvania,* articulating a Burkean conservative view of colonial rights certain to reinforce his resolve. A year earlier, Franklin had written the preface to its English edition. The two men were on converging paths.

"AS THE STAMP ACT IS AT LENGTH REPEAL'D," Franklin had written to his wife from London in 1766, "I am willing you should have a new

Gown, which you may suppose I did not send sooner, as I knew you would not like to be finer than your Neighbours, unless in a Gown of your own Spinning."[32] Now that he viewed the boycott as over, Franklin lavished British products on family members in America. "Take one thing with another," he wrote at this time to his sister in Boston, "and the World is a pretty good sort of World; and 'tis our Duty to make the best of it and be thankful."[33] Reflecting this sort of pragmatic optimism in light of the Stamp Act's repeal, Franklin all but ignored the continuing slights of Sugar Act duties, Quartering Act impositions, and Declaratory Act provocations.

In 1767, then, the Townshend Act came as a shock to him, just as it did for other colonists who (unlike Mason and his ilk) wanted to think the best of Britain. Admittedly, the Townshend duties fell on the right side of the bright line between internal excises and external imposts that Franklin had so artfully drawn to secure the Stamp Act's repeal. That made them no less objectionable to colonists like Washington who fixed the divide at raising revenue versus regulating trade.

At first, Franklin tried to downplay the significance of the Townshend Act, but *Letters from a Farmer* made the case against it on compelling constitutional grounds. Franklin again had to play catch-up, which he did in part by writing his preface to *Letters from a Farmer*. Many in England even surmised that he penned the book itself, so much did the anonymous tract reflect his evolving views. He was surprised to learn that its author was the pro-proprietor Philadelphia lawyer John Dickinson, a longtime foe of Franklin's political faction. A common enemy was uniting Americans behind a shared cause.

"The parliament unquestionably possesses a legal authority to *regulate* the trade of *Great-Britain,* and all her colonies," Dickinson wrote in his second letter. "The single question is," he asserted, "whether parliament can legally take money out of our pocket, without our consent. If they can, all our boasted liberty is but a sound, and nothing else."[34]

Franklin was moving beyond even this point. Who can fairly distinguish "between duties for regulation and those for revenue?" Franklin asked in a 1768 letter to his son, the royal governor of New Jersey. "The more I have thought and read on the subject the more I find myself confirmed in opinion, that no middle doctrine can be well maintained," he noted. "Something might be made of either of the extremes; that Parliament has a power to make *all laws* for us, or that it has a power to make *no laws* for us; and I think the arguments for the latter more numerous and weighty than those for the former."[35]

For Britain and its American colonies, Franklin began envisioning some sort of commonwealth of separate states under a single crown, such as Scotland and England before the 1707 Acts of Union or later emerged with Dominion status for Canada. Being more a practical than an abstract political thinker, Franklin conceived of this only as events forced the issue. "The Colonies originally were constituted distinct States," he wrote in 1770. "Our Assemblies with the King have true Legislative Authority."[36] By the end of 1773, Franklin was lecturing his son, "the parliament has no right to make any law whatever, binding on the colonies."[37]

In the meantime, from his perch as an assembly agent in London, Franklin launched a vigorous private lobbying and public writing campaign against the Townshend Act. Most of all, like Washington, he urged colonists to boycott British imports until Parliament repealed the duties. This gradually burned his bridges in London, where the tenor turned increasingly against what many in England widely viewed as the colonists' intransigence.

Matters only became worse after resistance to the Townshend Act in Boston led the British to dispatch a warship and troops to enforce compliance there. The result was bloodshed. On March 5, 1770, a band of hard-pressed soldiers, their backs to the wall, fired on a harassing mob of several hundred colonists, many of them young, killing five in what quickly became known in the colonies as the Boston Massacre. The soldiers had never received an order to shoot. The first did so after being hit by a thrown object. The others followed with an undisciplined volley into the crowd. Three Bostonians died on the spot; two more later from their wounds. Shots struck six additional protesters, permanently disabling one of them.

Although the news had not reached America, in response to the ongoing boycotts, Townshend's successor as chancellor of the exchequer, Lord North, had already engineered partial repeal of the Townshend Act. To maintain the principle of its sovereignty, however, Parliament retained the tax on tea. That, coupled with outrage over the Boston Massacre, kept the situation in the colonies fluid, with the Sons of Liberty urging a continued boycott everywhere.

Franklin and Washington remained firmly committed to the boycott. In a widely republished letter dated two weeks *after* partial repeal, Franklin wrote concerning nonimportation, "It appears to me, that if we do not now persist in this Measure till it has had its full Effect, it can never again be used on any future Occasion with the least prospect of Success, and that if we do persist another year, we shall never afterwards have occasion to use it."[38] He put it succinctly in an earlier dispatch: "It is not *the Sum* paid in that Duty on Tea that is Complain'd of as a Burthen, but the Principle of the Act."[39]

During the 1770 assembly session in Williamsburg, Washington served on a committee charged with revising the Virginia nonimportation association in response to partial repeal. He argued for maintaining the full boycott but, describing it as "the best that the friends to the cause coud obtain here," settled for a modified agreement that took some items off the proscribed list.[40] Branding it as "too much relaxd," Washington expressed his wish for one "ten times as strict."[41] Yet everywhere the boycott slackened until in many places it remained only on imported tea. Once moderates, Franklin and Washington had become hard-liners.

WITH PARTIAL REPEAL, relations between Britain and its American colonies lapsed into an eerily quiet period until late in 1773. Then, in coordination with Parliament's passing the Tea Act of 1773 (which lowered the price of imported tea without reducing the Townshend duty on it), the East India Company tried to break the boycott by flooding the American market with cheap tea sold directly to consumers. Alerted by the committees of correspondence, local chapters of the Sons of Liberty turned back ships in other ports, but Thomas Hutchinson, who was now the royal governor of Massachusetts, refused permission for those in Boston harbor to depart without off-loading their tea. The Company had twenty days to do so. Perhaps recalling the destruction of his home during the Stamp Act crisis, Hutchinson wanted to establish the principle of parliamentary sovereignty in the city most devoted to colonial rights. Protests organized by the Sons of Liberty prevented anyone from unloading the tea. A highly visible standoff ensued. Its consequences were as foreseeable as they were unforgettable.

Unlike the Boston Massacre, the event later known as the Boston Tea Party was a planned revolutionary act. Summoned by the Sons of Liberty, thousands of Bostonians—up to a third of the town's population—assembled on December 16, 1773, in and around a meeting hall near the port to make one last demand that the ships leave before the twenty-day deadline passed. Receiving no relief, later that day a hundred or so of the leaders, lightly disguised as American Mohawk warriors, boarded the ships in full view of armed guards and systematically disposed of 342 chests of tea by tossing their contents into the sea. Upon leaving, they swept the decks clean of loose tea leaves. Although they never publicly identified themselves, participants likely included such prominent Bostonians as John Hancock, Paul Revere, and Samuel Adams.

Parliament reacted to the Boston Tea Party with rage born of pent-up frustration. It passed legislation closing the port of Boston until colonists paid for the tea. Another new law shifted the locus of governing authority in Massachusetts from the elected assembly to the appointed governor. A third of the so-called Intolerable Acts empowered the governor to transfer trials of officials charged with capital offenses to courts in Britain, which Washington dubbed the Murder Act since he said it could allow those officers to get away with murder. A fourth act authorized royal governors to order the quartering of troops in vacant public buildings. Further, Britain replaced the ineffective Hutchinson as the Massachusetts governor with the commander in chief of British forces in North America, Thomas Gage. Gage soon moved the bulk of his troops to Boston, which he viewed as the rebel heart of America. "They will be lyons, whilst we are lambs; but, if we take the resolute part, they will undoubtedly prove very meek," Gage advised George III in February 1774. Four thousand British soldiers, Gage said, should be "sufficient to prevent any disturbance."[42]

If Parliament intended to intimidate other colonies by punishing Massachusetts, it failed. They rallied behind their sister colony, fearing that what happened to Massachusetts could await them. Even colonists like Franklin and Washington, who denounced the protesters for destroying private property, deplored

Britain for obliterating the property rights and traditional liberties of the many for the actions of a few. Punish those few, they asserted, not an entire city and colony.

Franklin was no longer in a position to have influence in London, however. A year earlier, he had forwarded to the speaker of the Massachusetts assembly old letters from Hutchinson to a high ministry official in London about how to handle the situation in Boston. "There must be an abridgment of what are called English liberties," Hutchinson had advised.[43] He even proposed punishing people for boycotting British goods. "Keep secret every thing I write," Hutchinson added.[44] Franklin never revealed who gave him the purloined letters, but their content was explosive. They suggested that the colony's governor was responsible for much of the oppression that the colonists had suffered. The resulting mistrust may have contributed to the impasse over tea.

Franklin's role in supplying the letters remained unknown for months, but it came to light just as news of the Boston Tea Party reached London. Now he was caught in the crosshairs of an enraged Parliament and royal administration. Critics piled on. By divulging ill-gotten private letters, they argued, Franklin had stirred the mob in Boston. He was subject to a brutal grilling before the Privy Council and lost his post as deputy postmaster for the colonies. Access to British officials ended and he could no longer function as a colonial agent. Some British friends remained loyal, but the time had come for Franklin to turn his sights toward home. He feared that war was now inevitable.

ALTHOUGH FRANKLIN AND WASHINGTON had similar disgust for the Intolerable Acts, Washington was better positioned to guide the colonial response. The Virginia House of Burgesses was in session when the news reached Williamsburg. It reacted swiftly. On May 24,

1774, the members passed an order condemning "the hostile Invasion of the City of *Boston* . . . by an Armed force" and calling for a day of fasting and prayer seeking divine aid "for averting the heavy Calamity which threatens destruction of our Civil Rights, and the Evils of Civil War."[45] Upon reading this order and assuming that more like it were on the way, Virginia's royal governor, Lord Dunmore, abruptly dissolved the assembly.

Washington and virtually all members present for the session reconvened the next day at Raleigh Tavern to debate and sign a nonimportation agreement covering tea and other East India Company products.[46] The assembly had already created a committee of correspondence led by its speaker, Peyton Randolph, to coordinate efforts with other colonies. It now issued letters calling for "the Appointment of Deputies from the several Colonies to meet annually in a general Congress" to forge a coordinated response to British policy.[47] Finally, on May 30, those burgesses still present, which included Washington, called an extralegal meeting of all members for August 1, a so-called Virginia Convention, to consider further action after each had time to collect the sense of his respective counties on how to proceed.

To prepare for the Virginia Convention, Washington called a meeting of Fairfax County freeholders for July 5, but heavy rains kept all but local Alexandria residents from attending. Those present voted to send relief supplies to Boston and named a committee to draft resolutions for a rescheduled meeting on July 18. George Mason largely took over drafting duties. He stayed at Mount Vernon on July 17 and rode with Washington to the second meeting. Washington chaired that meeting and presented the draft resolves. Once they passed, he carried them to the Virginia Convention.

Going far beyond condemning taxation without representation, the Fairfax Resolves painted a sinister picture of British tyranny. "There is a premeditated Design and System, formed and pursued by the British Ministry, to introduce an arbitrary Government into his Majesty's American Dominions," they declared. "The Act inflicting ministerial Vengeance upon the Town of Boston, and the two Bills lately brought into Parliament for abrogating the Charter of the Province of Massachusets Bay, and for the protection and Encouragement of Murderers in the said Province, are Part of the above mentioned iniquitous System."[48] As news of these resolves spread across Virginia and then to other colonies, Washington's name went with it. He became a revolutionary leader.

At the time, Washington's private rhetoric mirrored the Fairfax Resolves. "That Government is pursuing a regular Plan at the expence of Law & justice, to overthrow our Constitutional Rights & liberties," Washington wrote on July 20, 1774, to his pro-British neighbor Bryan Fairfax about the administration of Lord North, who by this time had become prime minister. "Shall we after this whine & cry for releif, when we have already tried it in vain?, or shall we supinely sit, and see one Provence after another fall a Sacrafice to Despotism?"[49] No, he wrote to Fairfax's half brother George William at roughly the same time, "We shall not suffer ourselves to be sacrificed by piecemeal [to a government that is] endeavouring by every piece of Art & despotism to fix the Shackles of Slavry upon us."[50] Washington went to the Virginia Convention breathing fire.

A rebel act from its call to order on August 1 to its adjournment eight days later, the Virginia Convention met in Williamsburg while Lord Dunmore was away. Claiming to represent their counties and constituents, the former members of a dismissed assembly—with two-thirds of those eligible in attendance—approved a sweeping agreement to boycott all British imports beginning in November and, "unless American Grievances are redressed before the 10th day of August, 1775," stop all exports of tobacco.[51] The convention declared tea a "detestable" drink due to its association with "the present Sufferings of our distressed Friends in the Town of Boston" and condemned Gage for his "odious and illegal" acts of martial law in Massachusetts.[52] It elected seven of its members, including Washington, to represent Virginia at a general congress of all colonies scheduled for Philadelphia during the following month. Unfounded reports circulated that Washington offered to raise a regiment of one thousand men and lead them in relief of Boston.

Meeting from September 4 to October 26, 1774, the First Continental Congress gave Washington a chance to strut on an intercolonial stage. Torn between conservatives like John Dickinson, who simply sought limits on Parliament's power to tax colonists, and radicals like Patrick Henry and the Adamses, Samuel and John, who hoped to assert a broad array of colonists' rights, the Congress mainly approved an intercolonial nonimportation agreement modeled on Virginia's most recent one and called a second congress for May 1775.[53] With his gift for engineering collective solutions to communal problems, had he been there, Franklin may have been able to extract greater results from the Congress. As it was, he could do little more than tout its nonimportation association in London. For his part, Washington said little in sessions but impressed many with the calm demeanor under fire that would become his trademark characteristic.

Washington's apparent tranquility was mainly on the surface. In letters from Philadelphia, he attacked the North ministry's "Systematic ascertion of an arbitrary power, deeply planned to overturn the Laws & Constitution" and warned "that more blood will be spilt on this occasion (if the Ministry are determined to push matters to extremity) than history has ever yet furnished instances of in the annals of North America."[54] Betraying his readiness to

fight, before leaving Philadelphia, Washington bought accessories
for his militia uniform and ordered a European military treatise.
With Franklin there by the time it began six months later, Washington would return to Philadelphia for the Second Continental
Congress.

EVENTS ON THE GROUND had dramatically changed the political land-
scape by the time the Second Continental Congress convened on
May 10, 1775. Virginia, for example, had begun arming for war.
Volunteer militia companies sprang up in various counties, with
five asking Washington to command them. In a clear case of taxation with representation, to pay for munitions, Fairfax County
levied a poll tax that Washington and Mason collected. When
Dunmore refused to call the assembly for fear of what it might
do, in March 1775, the members met again as an extralegal convention, this time beyond the governor's reach in Richmond.
When it hesitated to pass resolutions authorizing mobilization for
defense, Henry electrified the convention: "Is life so dear, or peace
so sweet, as to be purchased at the price of chains and slavery?
Forbid it, Almighty God! I know not what course others may take,
but as for me, give me liberty, or give me death!"[55]

Henry's words struck like a thunderclap. No one now shouted
him down with cries of treason, least of all the assembly's speaker
and convention's moderator, Peyton Randolph. The resolutions
passed, with Henry and Washington put on a planning committee to raise, arm, and discipline troops. Both were also returned
to the Continental Congress, this time with Washington placing
second in the overall vote, behind only Randolph and ahead of
Henry.

Other colonies were arming as well, particularly in New England, which faced the most immediate threat from British troops.

In Massachusetts, the local militia began operating independent of the governor as soon as Gage assumed that post and town meetings continued in defiance of the Intolerable Acts. When Gage dissolved the colony's assembly in October 1774, it removed to Concord and reformulated itself into a provincial congress with full power to levy taxes, purchase supplies, and maintain troops. Gage governed only in regions occupied by his troops, which were largely confined to Boston. They did not extend even as far as Cambridge, which (along with Concord) became somewhat of a revolutionary command and supply center.

Largely as a result of his own actions and choices, Franklin watched these developments from the sidelines in London. With his service as a colonial agent effectively over following his denouncement before the Privy Council in January 1774, once the immediate risk of arrest for trafficking in stolen letters passed, Franklin could have returned to Pennsylvania. If so, he would have attended the First Continental Congress in September and been present for his wife's death in December. But London had been his home for fifteen of the past seventeen years and, whether out of stubbornness or the belief that he could still do some good there, he hesitated to leave.

His son William urged him to return. "It seems your Popularity in this Country, whatever it may be on the other Side, is greatly beyond whatever it was," William wrote to his father in May from Philadelphia. "You may depend, when you return here, on being received with every Mark of Regard and Affection."[56] Indeed, colonists had protested his denouncement and saw him as a fellow victim of British tyranny, whereas in London, many vilified him as a chief provocateur of colonial resistance. As the rift widened, Franklin still hoped against hope to broker reconciliation between his two countries.

Even though Franklin duly noted in late September, "I have

seen no Minister since January, nor had the least Communication with them," he fairly added, "The generous and noble Friends of America in both Houses do indeed favour me with their Notice and Regard."[57] Franklin based his hopes for reconciliation on these private contacts and the prospect that nonimportation might bring down the North ministry. He also continued making the case for the colonies in articles published under pseudonyms in popular journals. Meetings with such leading dissident lords as Chatham, Howe, and Camden as well as with public intellectuals like Edmund Burke and Joseph Priestley gave Franklin some reason to believe that a compromise solution might still succeed. By early 1775, however, he realized that talking with such people was a waste of time. They could not speak for the government and Franklin could not speak for the colonies. Any compromise they brokered held no promise whatsoever as the sides for which they contended pulled ever further apart. The final straw was added on February 1, when the House of Lords rejected out of hand a comprehensive solution personally offered by none other than the legendary Lord Chatham, the former William Pitt, who made his proposal with a nod toward Franklin in the gallery. The lords, Franklin concluded, "Have scarce Discretion enough to govern a Herd of Swine," and "the *elected* House of Commons is no better."[58] All hope lost, Franklin sailed for Philadelphia on March 20, 1775.

While Franklin was at sea and Washington prepared to leave for the Second Continental Congress, the shots rang out that launched a revolution. To bolster his army's position in Boston, early on April 19, 1775, Gage dispatched seven hundred British troops on a seventeen-mile march to capture the militia armory in Concord. Forewarned, thousands of militiamen gathered to stop them. The first skirmish occurred at the commons in Lexington, where eight militiamen died. Another broke out at a bridge out-side Concord. By then, little remained in Concord for the British to capture. Their mission frustrated, a long line of redcoats headed back to Boston on a winding road lined with trees, stone walls, and buildings. Now the militiamen got their revenge as they fired in coordinated formations from protected positions. A survivor of Braddock's defeat, when he heard his officers' accounts, Gage must have recalled scenes from twenty years before—but this time colonists did the shooting. The British suffered 250 casualties compared with under 100 for the colonists.

"Unhappy it is though to reflect," Washington wrote after hearing the details, "the once happy and peaceful plains of America are either to be drenched with Blood, or Inhabited by Slaves. Sad alternative!"[59] By slaves, of course, Washington referred to white colonists subject to British tyranny, not the human chattel of white colonists.[60] A fifteen-thousand-person militia army soon bottled up Gage's much smaller force in Boston. This was the army that Washington would inherit.

A new, urgent reality greeted delegates as they arrived in Philadelphia for the Second Continental Congress. A war for colonial rights (if not for outright independence) had begun. Massachusetts delegate John Adams had toured the still blood-soaked battlefield before departing for Congress. Commenting later on his feelings, he recalled thinking, "The Die was cast, the Rubicon crossed."[61] Delegates wanted to hear Adams's full report when he reached Philadelphia, but Pennsylvanians reserved their most fervent welcomes for two other delegates.

Franklin's packet ship docked in Philadelphia on May 5, five days before Congress convened. Bells rang to mark the occasion. "Dr. Franklin is highly pleased to find us arming," a newspaper reported. "He thinks nothing else can save us from the most abject slavery."⁶² Pennsylvania added him to its congressional delegation on the following day—at age sixty-nine, he was the oldest delegate from any colony.

Washington reached Philadelphia on May 9 along with other southern delegates. A militia regiment met him outside the city and, with bands playing, escorted his carriage into town—the only delegate to receive such a martial welcome.

America's most experienced diplomat and its best-known soldier brought the skills that other delegates now knew were needed by Congress. After not seeing each other since Franklin sailed for London in 1757, they met soon after Washington arrived in Philadelphia, most likely at Franklin's home, where the senior statesman frequently entertained fellow delegates. They also inevitably dined together, cooperated in committee matters, and talked in the Assembly Room, where the Congress met. Entrusted with sometimes overlapping responsibilities exceeding those laid upon other delegates, they worked closely together until the war's end. For America, they became the two indispensable leaders for its revolution.

Book II

PARTNERS IN A REVOLUTION

Four

TAKING COMMAND

TO DISTINGUISH THEM FROM LATER LEFTIST REBELS, some historians characterize Franklin and Washington as reluctant revolutionaries with much to lose—conservatives who fought to preserve historic English rights rather than to gain or redistribute power. This characterization is simplistic. Although rooted in English tradition, the idealized rights that Franklin and Washington sought to preserve (or gain) had sprouted in such fundamentally different soil from that of feudal or early modern England as to make them a distinctly New World species.

The critical right to representation for determining taxation and internal lawmaking meant something profoundly different in the colonies, where most free, white, adult males elected members of their assemblies, than in England, where only a tiny fraction of such people could vote in elections for the House of Commons and none at all for anyone in the House of Lords (which needed to concur in legislation and often supplied the prime minister). In some more democratic colonies, like Pennsylvania, partisan politics drew on a broad electorate unlike anything that would emerge in Britain for more than a century. Even in aristocratic colonies like Virginia, where a nonpartisan elite dominated the assembly, candidates needed to secure support from a diverse spectrum of voters. Although generally successful, Franklin and Washington did not always win their assembly races and could not take victory for granted. In 1775, their concept of a popular right to representative government was without precedent. Indeed, only three years later, Franklin joined others in hailing America's War for Independence as "the greatest revolution the world ever saw."[1]

Although complaints about taxation without representation launched the American Revolution, Franklin, Washington, and other patriot leaders had a long list of grievances against Parliament. They were up in arms about the Intolerable Acts closing Boston harbor and imposing military rule in Massachusetts. The Proclamation of 1763 and the Quebec Act had deprived colonists of their power to exploit the frontier, which personally impacted Franklin and Washington. A host of oppressive mercantilist laws consolidated imperial power over colonial trade and commerce. Revolutions are often about the distribution of power, and by 1775 the colonies demanded more. One of Washington's modern biographers, Joseph Ellis, captured the Virginian's view of the American Revolution: "Essentially, he saw the conflict as a struggle for power in which the colonists, if victorious, destroyed British pretentions of superiority and won control over half of a continent."[2]

Private factors inevitably influence public actions. Like many later, more radical revolutionaries, Franklin and Washington held personal grudges against the corrupt exercise of excessive power by the governing regime, which in their case meant the British government and its officers. Franklin's stemmed from his shabby treatment as a colonial agent in London, particularly in his humiliation before the Privy Council, which turned him against British rule. Washington's dated back to his second-class status as a colonial military officer during the French and Indian War and his inability then to gain a regular army commission. If not yet irrevocably committed to American independence, both men went to Philadelphia for the Second Continental Congress in 1775 resolved to effect a fundamental transfer of rights and power within the British imperial system.

FOR HIS PART, Washington arrived wearing the crisp, new, blue uniform of the Fairfax Independent Company, one of the county militias he then commanded in defense of Virginia against the specter of British tyranny. A copper gorget, the symbolic relic of a medieval warrior's neck armor that still adorned an officer's uniform in most European armies, caught many eyes. Engraved with the motto of colonial Virginia *"En Dat Virginia Quartam"* ("Behold Virginia Makes the Fourth"), Washington's neckpiece signaled his colony's continued loyalty as the king's fourth (and first overseas) domain even as it turned against Parliament.

Some historians surmise that, by his apparel, Washington signaled his desire to serve as commander in chief of patriot forces. Such interpretations run counter to his solemn assertions to family and friends that he did *nothing* to solicit the post.[3] He hardly needed to since he was the obvious pick. Washington had more military experience than any other member of Congress. The only plausible alternative with more military experience, the former British colonel turned rebel Charles Lee, was unacceptable due to his foreign birth and personal eccentricities. John Adams called him "a queer Creature" who loved dogs more than people, which Lee (when he heard it) took as a compliment.[4] Artemas Ward, the Massachusetts militia general then commanding the patriot forces besieging Boston, lacked support even from his own state's leaders. From Adams on down, most of them favored Washington over Ward. The Virginian also had Franklin's full support and, with it, the backing of Pennsylvania. Rather than to signal an interest in the joint command, Washington may have worn his militia uniform simply to show his willingness to lead his colony's defense of liberty until called to a higher post. Such a pose suited him.

With armed conflict already imminent when it convened, the Second Continental Congress quickly turned to Washington for military advice by tapping him to lead virtually every committee that had anything to do with defense. This contrasted sharply with his experience as a delegate to the First Continental Congress only six months earlier. Then, with Congress still hoping to secure concessions peacefully through trade boycotts, Washington served on no committees and virtually disappeared from the historical record except for the odd comment of one fellow delegate that he "speaks very Modestly" and displays an "easy, Soldierlike air."[5] Now, with the colonies moving to a war footing following the battles of Lexington and Concord, it was Washington's soldierlike bearing that mattered most.

On May 15, in one of its first acts, with Boston harbor closed and many fearing that New York's was next, the Second Continental Congress named Washington to chair a committee charged with recommending ways to defend New York. Viewing it as a pivotal central colony, delegates saw the defense of New York as critical to the security of the whole. Entrusting the task to Washington signaled the delegates' faith in him.

Anticipating any invasion to come by sea, Washington's committee urged the colony to take up defensive posts around New York harbor. Drawing on memories of the French and Indian War, New Englanders also feared a land invasion from Canada through northern New York, and dispatched their own forces to guard against it. On May 18, Congress learned that, without much of a fight, militia forces from western New England had captured the lightly manned but heavily armed British forts at Ticonderoga and Crown Point, which had been built during the 1750s in northern

New York to deter invasions from French Canada. They would now serve that same purpose against invaders from British-ruled Canada.

In addition to chairing its committee on the defense of New York, until naming him to command patriot forces in mid-June, Congress tapped Washington to lead three more committees on military preparedness. The first addressed ways to supply colonies with gunpowder, the second projected the cost of a yearlong military campaign, and a third drafted rules and regulations governing a new "American continental army."[6] No member was in greater demand for committee service than Washington. No one's opinion mattered more on war-related matters. Everyone already looked to him as the military commander should the need arise.

With moderates still hoping for reconciliation with Britain, Congress initially balanced its war preparations with calls for peace. For this, delegates turned to Benjamin Franklin, their most seasoned diplomat. On June 3, Congress named him to a committee charged with drafting a petition asking the king to redress colonial grievances. At the time, many colonists blamed Parliament for their woes and envisioned some sort of settlement in which America could have its own representative government under the king. Franklin had spent the prior winter in London discussing terms for just such an arrangement, only to see them rejected by Parliament when offered by no less a figure than Lord Chatham. Now the resulting plea, called the Olive Branch Petition and largely drafted by John Dickinson, tried to resurrect Franklin's idea. When asked by delegates about the likely response from London, based on his own bitter experiences there, Franklin warned them "to be prepared for the worst," by which he meant war.[7]

The delegates did just that, with Franklin and Washington pressed into further service. On June 10, Congress asked Franklin, its sole scientist, to serve on a committee to devise ways to manufacture saltpeter for gunpowder. He also chaired a committee to reestablish postal services among the colonies, which led to his appointment as postmaster general. Then, on June 14, Congress ordered that ten companies of riflemen from the middle colonies join the fifteen thousand New England militiamen besieging Boston, with the combined force designated as "the Army of the United Colonies."[8]

A day later, the Congress unanimously elected George Washington as the army's commander in chief. As such, he would lead a Continental Army composed of soldiers typically recruited or conscripted by the states but paid by Congress and any local or state militia assigned to work with it.

To fund the war effort, Congress authorized the first of many issuances of paper money, with Franklin named to the committee charged with designing and printing the bills. His design featured his sketch of a chain with thirteen links circling the motto "We are one."[9] Congress also placed Franklin, its best-known author and editor, on a key committee assigned to draft a declaration, for publication by Washington when he took charge of the army, setting forth the causes for taking up arms.

"THO' I AM TRULY SENSITIVE of the high Honour done me," Washington told Congress in accepting his commission, "yet I feel great distress, for a consciousness that my abilities and military experience may not be equal to the extensive and important Trust."[10] An honest appraisal of his qualifications, he expressed similar concerns in letters to his wife, brother, brother-in-law, and stepson. Washington's experience leading colonial militias and temporary regiments could not prepare him for commanding a large but still forming volunteer force wholly lacking in training and supplies against the world's preeminent professional army, which was augmented by Hessian mercenaries and a navy with complete command of the seas. A virtuous cause and popular support only counted for so much. But who else was there? After all he had said and done, Washington felt honor bound to accept Congress's call despite his reservations.

"It was utterly out of my power to refuse this appointment without exposing my Character to such censures as would have re-flected dishonour on myself," Washington explained in a poignant letter to his wife. After urging armed resistance to British tyranny, he could not refuse to lead it when asked. "It has been a kind of destiny that has thrown me upon this Service," he wrote, and if Washington did not believe much else in the way of religion, he did believe in personal destiny and a disembodied providence. "I shall rely therefore, confidently, on that Providence which has heretofore preservd, & been bountiful to me," he assured her.[11]

Washington elaborated on his sense of duty in a long letter written a day later to his brother John. "The partiality of the Con-gress," he wrote, "joind to a political motive, really left me with-out a Choice."[12] The political motive clearly involved the desire of delegates to give a continental character to the resistance by choosing a native southerner—or at least someone not from the northeast—to lead an army composed mostly of New Englanders. For this role, the Virginian had no equal.

Precisely what Washington meant by Congress's partiality is less immediately clear. Certainly many delegates thought that he looked the part. Thomas Jefferson, who joined Congress in June, wrote privately that Washington's appearance "was fine, his stature exactly what one would wish, his deportment easy, erect, and noble."[13] John Adams's wife, Abigail, commented to her husband in July upon meeting the newly commissioned gen-eral, "You had prepared me to entertain a favorable opinion of him, but I thought the one half was not told me. Dignity with ease, and complacency, the Gentleman and Soldier look agreably blended in him."[14] The Philadelphia physician Benjamin Rush, who joined Congress in 1776, wrote to a friend about Washington, "You would distinguish him to be a general and a soldier from among ten thousand people. There is not a king in Europe who would not look like a valet de chambre by his side."[15]

In their letters, these correspondents stressed a further attribute that surely added to Congress's partiality toward Washington: his republican virtue. Jefferson, for example, noted, "His integrity was most pure, his justice the most inflexible I have ever known, no motives of interest or consanguinity, of friendship or hatred, being able to bias his decision."[16] Abigail Adams wrote, "Modesty marks every line and feture of his face."[17] Rush added, "His disinterestedness, his activity, his politeness . . . have captivated the hearts of the public."[18] These and other patriot leaders held to the Whig Party view that an all-consuming craving for power within the Tory Party in London posed an ongoing threat to individual liberty and property throughout the Empire. Written constitutions, bills of rights, and the separation of power among branches of government help to check the human proclivity toward self-interested power-seeking, Whig ideology maintained, but only the republican virtue of disinterested leaders could prevent tyranny. As every delegate knew, a covetous general like Oliver Cromwell could pose as grave a threat to liberty as a despotic king. For its general, Congress looked to someone who did not seek power and seemed eager to set it aside at war's end. Washington personified the citizen-soldier: an American Cato or Cincinnatus.

Since the First Continental Congress, fellow delegates had observed Washington's apparent modesty. "Eloquence in public Assemblies is not the surest road, to Fame and Preferment," John Adams wrote. "The Examples of Washington, Franklin and

Jefferson are enough to shew that Silence and reserve in public are more Efficacious than Argumentation or Oratory."[19] In particular, Adams praised Washington for leaving the room during debate over his nomination as commander in chief, moments after Adams formally nominated the Virginian for the post with Franklin's full support on June 15. A student of political theater, Franklin likely smiled at the scene of the large-framed, fully uniformed colonel slipping out of the chamber by a rear door. Franklin knew the vote would go well. Washington accepted the post a day later, with delegates hailing his humility. "I am called by the unanimous voice of the Colonies to the command of the Continental army," Washington wrote in a widely distributed public letter, "an honor I did not aspire to—an honor I was solicitous to avoid upon full conviction of my inadequacy to the importance of the service."[20] Connecticut delegate Silas Deane captured the general mood when he noted that Washington "Unites the bravery of the Solider, with the most consummate Modesty & Virtue."[21]

Nothing marked the ascension of Washington more than his offer to serve without pay. "As no pecuniary consideration could have tempted me to have accepted this arduous employment," he told Congress, "I do not wish to make any proffit from it."[22] Never mind that he possibly received more for expenses than he would have in pay. "There is something charming to me in the conduct of Washington," Adams wrote at the time, "that he would lay before us an exact account of his expenses, and not accept a shilling for pay."[23] Many others made similar comments about Washington's refusal of pay, with one widely republished article calling it "A most noble example."[24] Of course, Washington's wealth made it possible, and the same was true for Franklin, who donated his pay as postmaster general to disabled veterans. The two positioned themselves as models of republican virtue, standing above temptation.

Upon reaching New York in late June on his way to the front near Boston, Washington made another key republican affirmation, this time on behalf of himself and his officers. "When we assumed the Soldier, we did not lay aside the Citizen," he told the New York Provincial Congress, "& we shall sincerely rejoice with you in that happy Hour, when the Establishment of American Liberty on the most firm, & solid Foundations, shall enable us to return to our private Stations."[25] This was precisely the assurance that New York's revolutionary lawmakers asked him for in advance. His vow to retire at war's end was hailed across America. When he later fulfilled this commitment, Jefferson wrote, "The moderation & virtue of a single character has probably prevented this revolution from being closed as most others have been by a subversion of that liberty it was intended to establish."[26] His republican integrity intact, Washington was unsullied by power and became a model for emulation.[27]

Although the secret nature of the proceedings makes it difficult to reconstruct their interactions during the six weeks that Washington and Franklin served together in Congress (before the Virginian took command of the army), available evidence confirms that they attended all or most sessions, served on committees with overlapping jurisdiction, and shared a common political outlook. Committed to fight for colonial rights without being as white-hot for independence as some, they worked with all sides in Congress and continued to correspond with loyalists outside Congress. Although it was mainly the product of an odd-couple collaboration of the radical Jefferson and the conservative Dickinson, both Franklin and Washington could readily subscribe to the "Declaration Setting Forth the Causes and Necessity of Their Taking Up Arms," which Franklin's committee produced and Washington published following his arrival in Boston. "Our cause is just. Our union is perfect," the Declaration stated after relating those causes. "In defence of the freedom that is our birth-right . . . for the protection of our property . . . we have taken up arms. We shall lay them down when hostilities shall cease on the part of the aggressors, and . . . not before."[28] Independence was not yet at issue, though surely on everyone's mind.

WHILE STILL IN NEW YORK, Washington received news of a battle for Bunker Hill that was destined to shape the war in New England. Since the clash at Concord three months earlier, a jury-rigged patriot army now numbering around sixteen thousand had tightened its grip on a British force less than half its size bottled up in Boston, which stood isolated at the end of a long peninsula. Having received reinforcements for losses suffered in the retreat from Concord, including the arrival of Generals William Howe, Henry Clinton, and John Burgoyne, British commander Thomas Gage decided to counter the envelopment by seizing the strategic highlands across the water to the north and south of Boston.

Learning of Gage's plans, the patriots moved on the night of June 16 to occupy two hills on the Charlestown peninsula north of Boston. Responding the next day with a frontal attack on patriot positions, British forces were twice repulsed before taking the hills after defenders ran out of ammunition. The British had won, but at great cost—suffering more than a thousand casualties (including a hundred commissioned officers) to less than five hundred by the patriots—and never regained the offensive in New England, leaving Dorchester Heights across the water south of Boston unoccupied.

Reactions came swiftly, with Americans claiming a moral victory and London all but conceding a tactical defeat. "Britain, at the expense of three millions, has killed 150 Yankies this campaign, which is £20,000 a head; and at Bunker's Hill she gained a mile of ground," Franklin commented in letters designed for publication in Britain. "From these *data*," he asked readers to "calculate the time and expense necessary to kill us all, and conquer our whole territory."[29]

After reviewing Gage's report of the battle, the British high command replaced him with Howe, and the king issued a proclamation declaring elements of the colonies in open and avowed rebellion. Undermining moderates at the Continental Congress and in the colonies, this proclamation, issued on August 23, 1775, extended the rebels' rage against illegitimate parliamentary rule to also include arbitrary royal decrees. Delegates opposing revolution, such as Philadelphia lawyer John Dickinson, could no longer effectively argue that King George would bridge the gap between Parliament and his colonies. He was now the problem, just as George Mason had argued all along. Franklin had realized this while still in London; Mason had drawn Washington into this line of thinking before the Continental Congress began. Bringing the public on board was a necessary step toward revolution.

The carnage at and around Bunker Hill, particularly as it was portrayed by patriot propagandists, helped steel resistance to British rule. Hearing of the battle's outcome while on his way toward Boston, Washington gained faith that the troops he would command could defeat the British. "A few more such Victories would put an end to their army and the present contest," Washington boasted after reaching his command post at Cambridge, Massachusetts, in July.[30] Of more worry to Washington was word, coming from his cousin at Mount Vernon, that Virginia's royal governor had responded to the rebellion by offering freedom for slaves who left their masters to fight for the crown. "There is not a man of them, but woud leave us, if they believe'd they coud make there Escape," Washington was told about his own slaves. "Liberty is sweet."

After Bunker Hill, Washington yearned for bold strokes that could sweep the British from the field but would not have the means to engineer them until Franklin secured French support. The siege of Boston never provided one. Outnumbered two to one, the British stayed behind their defensive perimeter protected by water on three sides and the Royal Navy in Boston harbor. "The place indeed is naturally so defensible," Franklin wrote about his hometown and the British troops there, "that I think them in no danger."[31] Washington's generals vetoed plans to storm the city, including one calling for troops to rush the heavily fortified neck of land connecting it to the mainland and another with troops skating over the frozen Charles River during winter.

Instead, Washington waited ten months until his chief artillery officer, Henry Knox, managed to have men under his command haul heavy cannons seized at Fort Ticonderoga over frozen mountain roads in the dead of winter and place them on the heights south of Boston. From there, these guns could destroy the city and any army in it. Without a fight, Howe evacuated his troops from Boston by ship on March 17, 1776. Rather than retreat to England, however, Howe's army reassembled in Halifax with reinforcements drawn from across the Empire and some eighteen thousand Hessian mercenaries for a massive assault on New York that tested Washington to his limits and drew Franklin to the front.

Franklin's role in the war effort had increased even before the siege of Boston ended. He turned seventy during this period and doubted whether he would survive the war, yet he took on a staggering number of tasks for Congress and his colony.[32] Beginning in July, he presided over Pennsylvania's Committee of Safety, a demanding job that involved organizing and managing the colony's defense. Between this committee's broad mandate and the virtual removal of the colony's appointed governor, Franklin effectively acted as Pennsylvania's chief executive. Regaining his seat in the colonial assembly, he also served on committees to enforce the boycott with Britain even as he championed efforts to allow exports in exchange for arms or ammunition and to open foreign markets for American goods. In August, he brokered a deal with merchants from Bermuda to trade stolen British gunpowder for Pennsylvania produce. Congress soon elected him to two powerful new secret standing committees: one to secure war matériel from abroad and one dealing with foreign affairs. It also named him as an Indian commissioner and to chair or serve on more than a half dozen ad hoc committees.

"I am here immers'd in so much Business that I have scarce time to eat or sleep," Franklin wrote. "My time has never been more fully employed."[33] Because Franklin had recently returned from London where he socialized with British officials and his son still served as New Jersey's royal governor, some delegates initially doubted Franklin's commitment to the patriot cause, but they were soon won over by his zeal. "He does not hesitate at our boldest Measures," John Adams wrote in July, "but rather seems to think us, too irresolute, and backward."[34] And no delegate had a wider range of expertise and experience to bring to the effort.

Franklin's experience led him to foresee sooner than most the need for undertaking two fundamental reforms that seemingly ran counter to the libertarian impulse driving the Revolution. First, he consistently spoke and thought in terms of "America" rather than of colonies, believing that the British could be defeated and liberty secured only by a "united" effort.[35] Having pushed the Albany Plan of Union during the French and Indian War, Franklin introduced a similar though somewhat stronger draft constitution for the colonies in July. Knowing that Congress was not yet ready for such a step, Franklin simply offered his proposal and asked that it be tabled for later consideration. Clearly federal in nature, with each colony retaining control over matters peculiar to it, Franklin's draft contained concepts that he would support at the Constitutional Convention, including proportional representation in Congress and centralized power over commerce, war and peace, foreign affairs, western lands, and such domestic matters as thought "necessary to the General Welfare."[36] These would become the same issues that Washington also embraced at the Convention.

Second, while many patriots, blinded by faith in their cause, believed that citizen-soldiers could quickly vanquish a hireling army, Franklin, sobered by experience dealing with obstinate British leaders, foresaw a long war. At its outset in May 1775, he predicted to David Hartley, a forty-five-year-old member of Parliament, that neither of them "may live to see" its end.[37]

Summer soldiers and part-time militias could not win such a war; only a disciplined, unified force could. Yet as commander of troops besieging Boston, not only did Washington face the daunting task of transforming volunteer militiamen from various colonies into a single, Continental Army, but he had to do so knowing that most of his men had signed up only to the end of 1775. Confronting an army of professional soldiers led by generals content to wait for the patriot cause to flag, Franklin and Washington knew that their army must be reconstituted on a more permanent basis. As Congress came around to this view once militia commissions began expiring with the British still embedded in Boston, it assigned Franklin to multiple committees charged with reforming the army. This work brought him back in direct contact with Washington on a common cause.

FRANKLIN HAD WRITTEN TO WASHINGTON only once since the general left Philadelphia in June. That came in August, when state militia under Franklin's command in Pennsylvania captured a British vessel with military supplies and army officers. Franklin sent them to Washington. By September, however, Washington faced a crisis demanding immediate attention. "My Situation is inexpressibly distressing, to see the Winter, fast approaching upon a naked Army: The Time of their Service within a few Weeks of expiring, & no Provision, yet made for such important Events," he wrote to Congress on September 21. "A Dissolution of the present Army therefore will take Place, unless some early Provision is made agst such an Event."[38]

Congress responded by sending a three-member delegation chaired by Franklin to meet with Washington and leaders from the New England colonies at army headquarters in Cambridge. There, Franklin stayed with Washington in a stately, three-story Georgian mansion commandeered from the loyalist brother-in-law of the colony's royal lieutenant governor. The house stood only a short walk from Harvard College, which had awarded Franklin its first-ever honorary degree two decades earlier, and Cambridge Commons, where Washington drilled the troops. The meeting of Washington, Franklin, and the others produced the framework for a new Continental Army.

The conference lasted for a week and addressed a range of issues relating to transforming a mishmash of militias jealous of their privileges into a Continental Army free of provincial distinctions. By all accounts, Washington and Franklin saw eye to eye and got their way—though executing their ambitious plan proved difficult on Congress's limited budget. Upon taking command, Washington had ordered "that exact discipline be observed, and due Subordination prevail thro' the whole Army,"[39] but he never fully achieved either with existing recruits. These principles ran through conference recommendations for the new army. Pay, rations, rank, and arms would be standardized on stated terms. Clothing and blankets would be supplied to soldiers who lacked them. The death penalty was authorized for treason and mutiny; lashings for soldiers convicted of insubordination and a host of other offenses. Officers would be cashiered for cowardice, fraud, or drunkenness.

Upon taking command, Washington said of the old army, "The Men would fight very well (if properly Officered)" but "their Officers generally speaking are the most indifferent kind of People I ever saw."[40] He sought to set this right in composing the new army by retaining the better officers for reorganized units and letting the others go. Overall, the conferees agreed to expand the army's size to roughly twenty thousand officers and men—an ambitious target it never reached. State regiments and militia units, which sometimes worked in coordination with the Continental Army but also performed local defense duties under the command of their governor, always competed with the regular army for soldiers and officers.

Washington saved one issue to discuss with the delegates from Congress after the conference ended and the governors left. For the new army, he did not want to recruit slaves or freed Blacks, or even to reenlist Blacks already serving in existing units, as many did in the Massachusetts and Rhode Island militias. Washington took slaves from Virginia to the front, but only as personal servants wearing the livery of Mount Vernon, not in military uniform. It is unclear how Franklin responded to Washington's request. By this point, the Pennsylvanian had long held that Blacks were innately as capable as whites. Indeed, twelve years earlier, after visiting a school in Philadelphia for Black children, Franklin reported finding the students' abilities equal to that of whites.[41] The other two delegates from Congress were southern slaveholders, however, and the delegates informally agreed to Washington's request without including it among their official recommendations.[42] Recruiters initially followed the new policy, but manpower needs soon overwhelmed it. Within about a year, free Blacks serving alongside whites constituted a tenth of the Continental Army.

FRANKLIN USED THE OPPORTUNITY of this trip to reunite with his sister Jane, who had been driven from her home in Boston by the British occupation, and take her to Philadelphia with him. This reunion roughly coincided with his complete estrangement from his loyalist son William, who was soon forced out as governor of New Jersey, arrested, exchanged for a captured patriot governor, and eventually resettled in England. Once so close as to present a model father-son relationship, the two never reconciled. Franklin could not forgive his son's allegiance to the crown.

As Franklin headed south from Massachusetts with his sister, Martha Washington prepared to head north to join her husband for the winter encampment. Such visits became a regular feature of their wartime marriage. The Revolution utterly disordered domestic life for both families, as it did so many others. After leaving the estate in 1775, Washington did not see Mount Vernon again until 1781, and then only briefly. Although initially able to remain mainly in Philadelphia, Franklin ultimately spent more years away from his home for wartime service than Washington.

Indeed, after returning from Massachusetts in November 1775, Franklin was off again on an arduous trip to Montreal the following March. Hoping disaffected Canadians would rally to the revolutionary cause, Congress had ordered, and Washington had supported, an ill-conceived, ill-planned, and ill-executed invasion of Canada in 1775. It captured Montreal and much of the upper St. Lawrence Valley in November but stalled before the fortress at Quebec City in December. With the British unable to mount a counterattack with fresh troops until spring, the Americans settled into an uncomfortable occupation of the regions they had captured. While those occupying Montreal initially enjoyed some local support, it waned as they began suppressing loyalists and became unable to pay for requisitioned supplies. Pro-British landowners and clerics also warned that the invaders would impose their Protestant religion on the colony's largely Catholic population.

Suddenly the liberation promised in a published address by Washington to the Canadian people did not sound so liberating. "Come," he had written at the invasion's rosy outset, "unite with us in an indissoluble Union, let us run together to the same Goal. We have taken up Arms in Defence of our Liberty, our Property, our Wives, and our Children. We are determined to preserve them, or die."[43] For French Canadians only recently subjected to British rule, American promises of freedom were unpersuasive.

In the spring of 1776, then, in a remedial measure almost as ill conceived as the initial invasion, Congress dispatched a commission led by Franklin to revive local support—even as Britain massed troops in the lower St. Lawrence Valley to relieve Quebec and retake Montreal.[44] Although Congress had the good sense to name a Catholic to the commission and, in Franklin, someone committed to religious freedom and with some fluency in French, it did not provide commissioners with the hard currency, supplies, or troops needed to win back local trust. Instead, it carried copies of Common Sense, patriot propagandist Thomas Paine's new pamphlet, which was working wonders in turning Americans against King George but had less effect in Canada.[45]

"I am here on my way to Canada, detain'd by the present state of the Lakes, in which the unthaw'd Ice obstructs Navigation," Franklin wrote from upstate New York in April. "I begin to apprehend that I have undertaken a Fatigue that at my Time of Life may prove too much for me."[46] Heavy snow blocked overland routes. Franklin had the foresight to bring his own bedding to guard against fleas and lice, but still returned with boils, rashes, and infections. On the way north, commissioners slept on the bare floors of war-ravaged houses. In Montreal, creditors hounded them to pay for what the troops had taken. Dire reports flooding in from the front near Quebec City spoke of a dwindling American army—ravaged by smallpox, reduced by expiring commissions, and relying on scant supplies—facing the imminent threat of fresh British forces arriving by sea.

"It is impossible to give you a just idea of the lowness of the Continental credit here," Franklin wrote to Congress from Montreal on May 1. "The general apprehension, that we shall be driven out of the Province as soon as the King's troops can arrive, concurs with the frequent breaches of promise the Inhabitants have experienced, in determining them to trust our people no farther."[47] A week later, the commission advised Congress, "If Money cannot be had to support your Army here with Honor, so as to be respected instead of being hated by the people, . . . it is better immediately to withdraw it."[48] Although the news had not yet reached Montreal, by this time American forces were already pulling back from the region around Quebec City in the face of British advances. The withdrawal soon turned into a panicked retreat as the British pushed up the St. Lawrence Valley. Learning of the rout, Franklin abandoned Montreal for the safety of upstate New York on May 12, with the remaining American troops following in mid-June.

For all its failings, the invasion of Canada shed light on the revolutionary vision of Franklin, Washington, and other patriot leaders. Believing that people want freedom, they thought Canadians would rally to the cause if given a chance: only force and misinformation kept them in chains. Washington foresaw Canadians greeting his troops as liberators. "Range yourselves under the Standard of general Liberty," he urged.[49] Evidence forwarded by Franklin's foreign affairs committee to Congress leading to the commission's appointment spoke of lies by local Catholic clerics and nobles turning Canadians against "the American side."[50] Congress instructed the commission to counter those lies. America invaded "for the Purpose of frustrating the Designs of the British Court against our Common Liberties," the commissioners were told to proclaim. "Declare that it is our Inclination that the People of Canada may set up such a Form of Government, as will be most likely, in their Judgment, to produce their Happiness," Congress asserted. "Declare that we hold sacred the Rights of Conscience, and may promise to the whole People solemnly, in our Name, the free and undisturbed Exercise of their Religion." To get out this word, Congress directed the commission, "You are to establish a free Press."[51]

Issued four months before the Declaration of Independence, these instructions trumpeted America's founding principles: liberty for all, republican self-government, freedom of conscience, and a free press. At a time when most colonists spoke English and viewed Catholicism as an alien religion, Congress directed the commissioners to assure the people of Quebec, "It is our earnest Desire to adopt them into our Union as a Sister Colony, and to secure the same general System of mild and equal Laws for them and for ourselves, with only such local Differences, as may be agreeable to each Colony respectively."[52]

The United Colonies, as Congress still called the American union, was to be federal in nature, with common core principles married alongside provincial variations, and open to ethnic and religious diversity.[53] Washington stressed this in his address to the Canadian people: "The Cause of America, and of Liberty, is the Cause of every virtuous American Citizen; whatever may be his Religion or his Descent," he wrote.[54] At a time when many colonies still supported an established Protestant church and imposed limits on Catholics, after assuring Canadians of the right to practice their Catholic religion, Congress's instructions went on to provide that members of "other Denominations of Christians" in Quebec be equally entitled to practice theirs and "be totally exempt from the Payment of any Tythes or Taxes for the Support of any religion."[55] Here was a strong free-exercise right along with antiestablishment principles. With Franklin and Washington yoked equally to the effort, Congress was defining what it meant to be American.

AS AMERICAN TROOPS RETREATED from Canada during June 1776 in the face of a British army, a second and much larger British army sailed toward New York harbor—a show of strength designed to end the rebellion once and for all. Anticipating the move, Washington had shifted his main force from New England to New York in April and begun readying harbor defenses. In light of the expected attack and before the final retreat from Canada, Congress called Washington to Philadelphia for talks that extended from May 23 to June 3. Before leaving for them, he wrote to Franklin about the debacle in Canada. "Tho your presence may conduce to the public good in an essential manner," Washington said of Franklin's labors in Montreal, "yet I am certain you must experience difficulties and embarrasments of a peculiar nature. Perhaps in a little time, Things may assume a more promising appearance than the present is."[56]

After a hard trip home that left him sick for weeks, Franklin arrived in Philadelphia on May 30 with the latest news of defeat in Canada. "The ignorance of the Canadans, there Incapasity and Aversnes to have any thing to do with war and his Indisposition I believe Affected His Spirets," Franklin's sister observed.[57] That indisposition, which Franklin attributed to a recurrence of gout, lasted nearly a month. "I find I grow daily more feeble," Franklin complained in late May.[58] His first days back were active, however, as he caught up on his work at Congress and discussed the situation in Canada and New York with Washington, before taking to his bed after the general left. By various accounts, their discussions were somber but not desperate. Indeed, after them, Franklin wrote with his characteristic optimism to Washington about British advances in Canada and on New York, "I see more certainly the Ruin of Britain if she persists in such expensive distant Expeditions."[59] Franklin would soon be at the front again, but not before taking center stage in the Revolution's signature event. On this too, Washington stood in full accord.

ON MAY 15, 1776, in light of the British offensive, the Virginia Convention had voted to direct its delegation in Congress to offer a motion for independence. These instructions arrived before Washington left Philadelphia. "I am very glad to find that the Virginia Convention have passed so noble a vote," Washington wrote at the time. "Things have come to that pass now, as to convince us, that we have nothing more to expect from the justice of G: Britain."[60] Franklin was out of town recovering from his fit of gout on June 7, when Virginia delegate Richard Henry Lee formally offered the motion in Congress, but was nevertheless appointed to a "Committee of Five" to draft a declaration to accompany it. Jefferson, Adams, Robert Livingston, and Roger Sherman—some of Congress's best thinkers and writers—joined Franklin. Congress delayed the vote on independence until early July so that delegates could receive instructions from their colonies on how to vote, giving time for the committee to do its work. Jefferson assumed lead writing duties but submitted his draft to Franklin, Adams, and the full committee for review.

A seasoned editor, Franklin knew good copy when he saw it and made few changes to Jefferson's draft. Britain's "arbitrary power" became its "absolute Despotism." The charge of Britain "taking away our Charters, & altering fundamentally the Forms of our Governments" expanded to also include "abolishing our most valuable Laws." The colonists' petitions being "answered by repeated injury" hardened to being "answered only by repeated Injury." And most profoundly, for the foundation of human rights and equality, "We hold these truths to be sacred & undeniable," rose to the phrase that rings down through history, "We hold these truths to be self-evident." Reason trumped revelation. Other edits sharpened the long list of specific charges levied against the king.[61]

Franklin made his edits from his sickbed in the country but returned to Congress in time for debate on the document, which began on June 28. Congress reworked segments and struck whole paragraphs on political grounds, including most passages critical of slavery. The process visibly irritated Jefferson. Franklin attempted to comfort (or at least distract) him with an anecdote. As Jefferson later related Franklin's story, after composing a sign for his shop that read "John Thompson, *Hatter, makes* and *sells hats* for ready money," joined with the picture of a hat, the hatmaker showed it to his friends. Each offered a different redaction until the sign included only the shopkeeper's name "with the figure of a hat subjoined."[62] At least Jefferson's Declaration of Independence fared better than Thompson's sign.

Congress approved the revised Declaration on July 4, 1776, two days after Lee's resolution, and it immediately became a rallying cry for revolution. Washington had it read to the troops on July 9, prefaced with his commentary that it should "serve as a fresh incentive to every officer, and soldier, to act with Fidelity and Courage, as knowing that now the peace and safety of his Country depends (under God) solely on the success of our arms."[63] When Congress's president John Hancock signed the parchment copy with a stern warning to fellow delegates, "We must all hang together," Franklin reportedly added the pithy retort, "or most assuredly we shall all hang separately."[64]

BY THEN, the American cause needed whatever boost it could get from the declaration. On July 2, General Howe began landing troops on undefended Staten Island—across New York harbor from Manhattan—amassing an invasion force that exceeded twenty-five thousand men by early August. Washington's chief lieutenant, Henry Knox, depicted local patriots as "frigh'd to death," while Washington worried about the encouragement given to loyalists.[65] "We expect a very bloody Summer of it at New York," he wrote, "as it is there I expect the grand efforts of the Enemy will be aim'd."[66] Despite an order from Congress directing him to defend the city, its extended coastline and open terrain made it all but indefensible against a seaborne invasion by an overwhelming force.

Despite their military advantage, Howe and his brother, the operation's naval commander Admiral Lord Howe, still hoped for peace and carried commissions empowering them to offer generous terms to secure it: pardons for all and restored rights for colonies. This was the same Lord Howe who, as an independent member of the House of Lords, had tried to negotiate privately with Franklin in London along similar lines, only to have Parliament reject any limit on its right to tax the colonies. Now, after lobbying the king, he had enhanced authority to bargain for peace.

As his brother assembled the army on Staten Island, Admiral Howe sent circuitously worded letters to Washington and Franklin inviting them "to converse" on "the reestablishment of lasting Peace and Union with the Colonies."[67] Not authorized to acknowledge their official status under a rebel government, Admiral Howe addressed both men as private citizens. Washington refused to receive Howe's letter to him because it did not address him as general. Franklin, who was a private citizen as well as a delegate, turned his over to Congress and dispatched a cold reply to Howe. If *"Peace* is here meant," Franklin wrote, it must be "a Peace between Britain and America as distinct States now at War."[68] This Admiral Howe had no authority or inclination to discuss. There things stood until a series of deft British maneuvers against American positions on Long Island and Brooklyn Heights forced Washington's reeling army back onto Manhattan island, where it risked entrapment. So far in the New York campaign, the Americans had been outmanned, outgunned, and outgeneraled at every turn, with worse to come.

With his brother pausing the assault at a point when it might have crushed patriot resistance, Admiral Howe again reached out to Franklin in a letter sent by way of Washington. While noting that he was not empowered "to negociate a reunion with America under any other description than as subject to the crown of Great Britain," Howe now made clear that he was authorized to discuss terms like those laid out by Congress a year earlier in its Olive Branch Petition, which suggested that America and Britain have separate assemblies under a single king.[69]

"The time is past," Franklin wrote in a hastily drafted but never posted reply. "One might as well propose it to France."[70] On reflection, he waited to hear what Congress wished in light of the situation in New York.

Admiral Howe also sent a captured American officer named John Sullivan to press Congress for the conference. As directed by his captors, Sullivan claimed that the Howes had "full power to Compromise the Dispute between Great Britain and America, upon Terms advantageous to both."[71] John Adams dismissed Sullivan as a decoy, but Congress decided to authorize a meeting anyway, with Adams and Edward Rutledge accompanying

Franklin. By dispatching letters back and forth between the sides, Washington served as intermediary for the two parties, which met on September 11 at Staten Island's southern tip, across from American-held New Jersey.

Scarcely recovered from his mission to Canada, Franklin pushed his way through New Jersey on roads clogged with refugees and soldiers. The lack of available accommodations on the outbound journey forced him to share a bed with Adams in a small room with a single window that Franklin (who insisted on fresh air when sleeping) wanted open and Adams (who feared a cold) wanted shut. No one catches colds from cold air, Franklin insisted; they spread from person to person in stagnant air. He likely eyed the already ailing Adams as a possible source of contagion. "The Doctor then began an harrangue, upon Air and cold and Respiration and Perspiration, with which I was so much amused that I soon fell asleep, and left him and his Philosophy together," Adams later recalled.[72]

The conference itself was a bust. Admiral Howe opened with a long discourse on his personal affection for America and the king's desire to grant generous reforms. But, according to the report submitted by Franklin's committee to Congress, Howe offered "no explicit proposition of peace" without the Americans' submission to British authority. To this, the report noted, "We gave it as our opinion to his lordship that a return to the domination of great Britain was not now to be expected" in light of Britain's past failures to address colonial grievances and popular support for independence.[73] This effectively ended the meeting, although Adams later reported Howe's adding that he would feel and lament America's fall "like the Loss of a Brother." In response, Franklin, with "a Bow, a Smile and all that Naivetee which sometimes appeared in his Conversation and is often observed in his Writings, replied 'My Lord, We will do our Utmost Endeavours,

to save your Lordship that mortification.'"[74] Howe's personal secretary captured the meeting's tenor in a journal entry. "They met,
they talked, they parted," he noted of the four principals, "and
now, nothing remains but to fight it out."[75]

Franklin's return to Philadelphia marked the end of his visits
to the front and a transition in his relationship with Washington.
Since its formation in 1775, Franklin had served on Congress's secret committee dealing with foreign affairs and, since June, on
one openly planning treaties and alliances with foreign powers.
This work, plus his experience as a colonial agent in London and
international renown, made him the logical pick to lead the new
nation's diplomatic offensive in Europe.

ON SEPTEMBER 26, 1776, Congress elected Franklin as a commissioner
to France, to serve with two American agents already in Europe,
Silas Deane and Arthur Lee, to secure foreign support for the Revolution. With the Howes resuming the attack in mid-September
and quickly driving Washington's army out of New York, Franklin's efforts in France became as vital for the American cause as
anything happening on the battlefield. France was Europe's other
great power and England's historic rival. If France did not help
America, no other country would. Franklin and Washington
began working together at a distance, with the success of each
dependent on that of the other. America needed French money,
arms, troops, and naval support, which France would supply only
if the American army showed some chance of winning the war.

As important as it was, Franklin's foreign assignment took him
away from his work for Pennsylvania and in Congress, responsibilities which had grown since the Declaration of Independence.
Beginning in mid-July, Franklin presided over the convention
drafting a new constitution for a now independent common-

wealth of Pennsylvania. Most other states engaged in similar efforts. Inspired by new notions of liberty (and no longer bound to the crown), they sensed a pressing need to replace their old royal or proprietorship charters with republican constitutions. Under Franklin's leadership, Pennsylvania crafted the most democratic of all the new state charters. It featured broad voting rights for free males, a unicameral legislature, recallable judges, an executive council presided over by a weak president, and an expansive declaration of rights. After decades of proprietorship control, the people would rule in Pennsylvania.

Simultaneously and in the same historic building, Franklin participated in congressional debates over articles of confederation for the newly independent United States. Without unity under the crown, the newly sovereign states needed to forge their own working alliance. In these debates, Franklin argued for allocating representation in Congress on the basis of population rather than giving each state one vote and, as a means to discourage slavery, counting slaves as people for purposes of computing the per capita financial contribution of each state. Delegates from southern states naturally wanted to count only free people to determine their state's share. Called from these debates for service in France, Franklin had staked out positions he would support twelve years later at the Constitutional Convention. Although war service prevented him from participating in such efforts for his state or in Congress, Washington recognized their vital importance. "To form a new Government," he wrote at this time, "requires infinite care, & unbounded attention; for if the foundation is badly laid the superstructure must be bad."[76]

Franklin left for France in October aboard the *Reprisal*, a fast but unsteady brig acquired by Congress to wreak havoc on British merchant ships and, if possible, capture them for prize money. For company, he took along his grandsons Temple, who served as his secretary, and Benny, who would receive a proper European education. Slipping through the British naval blockade of American ports, the *Reprisal* crossed the Atlantic in one seasickness-inducing month, snagging two British vessels on the way. Later admitting the voyage "almost demolish'd me," Franklin disembarked at first sight of land and traveled by carriage to Paris.[77]

The French greeted Franklin with acclaim. To them, he was both an Enlightenment philosopher like Voltaire and the personification of Rousseau's freedom-loving natural man, plus a renowned scientist and England's enemy to boot. Franklin became the man of the hour and person of the age in Paris, with medallions struck bearing his image. "These," he wrote to his daughter, "with the pictures, busts, and prints, (of which copies upon copies are spread every where) have made your father's face as well known as that of the moon."[78] Knowing the value of image in fashion-conscious France, he played the rustic sage, with uncoiffed hair, frontier fur cap, and bifocals. "Think how I must appear among the Powder'd Heads of Paris," he wrote.[79]

Franklin became a fixture in the finest salons of the French capital and resumed his scientific studies even as he served as America's senior diplomat in Europe. Ladies of the French court particularly favored him, and he them, which gave Franklin access to the inner workings of pre-revolutionary French society. French philosophes welcomed him too, and made him one of their own. These activities complemented one another to reinforce Franklin's already legendary stature. Born into a working-class family on the edge of civilization, Franklin's entrée into aristocratic high society symbolized the promise of America.

Winning over French society strengthened Franklin's hand but, so long as the American cause looked as desperate as it did after the battles in New York, gaining French support would take more than style. Franklin gave substance to American foreign policy by coupling cold realism with glittering idealism. Using the simple calculus that helping America weakened Britain, Comte de Vergennes, the French foreign minister, signaled a desire to aid the rebellion even before Franklin arrived. With France still rebuilding its navy after the Seven Years' War, however, it dare not risk war with Britain. Instead, it supplied secret support to America in the form of loans, trade, and (in some cases) covertly opening French ports to American privateers.

From the outset, Franklin pressed for more, with his commission's initial memo to Vergennes requesting ships, arms, artillery, ammunition, and a military alliance with France and its ally, Spain. "The Interest of the three Nations is the same," the memo explained: weakening Britain and expanding commerce. Supporting America's War for Independence would advance both objectives and perhaps result in France and Spain securing British colonies in the West Indies. Without aid, however, "our People" would be forced to reach some "accommodation" with Britain, the memo added.[80] This became Franklin's refrain: help us or else.

But he honed other arguments as well. Franklin never tired of telling Europeans that America was fighting for freedom, not self-aggrandizement. While the idealistic aspects of Franklin's diplomacy may not have appealed to the interests of the French court (which naturally feared republicanism), it carried weight among those who could influence French policy. Franklin knew this and used it. After his arrival in France, for example, he arranged for translating and publishing America's republican state constitutions and Articles of Confederation in Paris. "All Europe is for us," Franklin's commission reported in a March 1777 memo informing Congress of its activities. "The Prospect of an Asylum in America for those who love Liberty gives general Joy, and our Cause is esteem'd the Cause of all Mankind."[81] Recognizing the practical value of popular goodwill, Franklin instinctively linked realism and idealism in the pursuit of national interests. "He knew that America had a unique and powerful meaning for enlightened reformers in France, and that he himself, his very existence, was the embodiment, the palpable expression, of that meaning," historian Bernard Bailyn observed.[82] That meaning was Franklin's greatest asset.

ALTHOUGH FRANKLIN LAID THE FOUNDATION for an alliance with France, only the prospect of success on the battlefield could secure it. That prospect remained dim and distant during Franklin's initial year in Paris. First came the disastrous New York campaign that drove Washington's army, or what was left of it, out of Manhattan, across New Jersey, and into Pennsylvania by early December 1776. With units of General Howe's army having reached the Delaware River by then, Washington warned Congress on the ninth about Philadelphia, "The Enemy might . . . march directly in and take Possession."[83] He thought that Howe would take the city as soon as the river froze enough for his troops to cross over it. "In truth, I do not see what is to prevent him," Washington wrote.[84] Congress decamped for Baltimore. Meanwhile, a British force under Henry Clinton took and held the strategic harbor at Newport, Rhode Island. Between captures, casualties, disease, and desertion, Washington's army had dwindled to scarcely a few thousand soldiers fit for duty, with the terms of enlistment for most of these remaining men due to expire on December 31. "In a word," Washington declared, "if every nerve is not straind to recruit the New Army with all possible Expedition I think the game is pretty near up."[85]

The impending deadline forced Washington to gamble. On Christmas night 1776, scarcely a week before most of his men were free to leave, in a desperate effort to restore morale and regain the initiative, Washington took this army back across the ice-choked Delaware River and captured the Hessian garrison at Trenton, New Jersey. Buying more time by paying each man who would take it a ten-dollar bounty for an extra six weeks' service, Washington held off a British counterattack at nearby Assunpink Creek and routed a British force in Princeton before retiring with his remaining troops to Morristown, New Jersey, for the winter.

Dismissing the episode as a minor setback, General Howe pulled back most of his troops to New York for the winter. At the time, European armies typically spent the winter in quarters, a custom both sides followed throughout the American Revolution. Philadelphia was safe for another year. Congress returned to the city.

While these pivotal victories boosted American spirits, they could not dissipate the gathering gloom as soldiers left the Continental Army in droves once their commissions expired. After pulling in men from other units, Washington encamped for the winter in Morristown with about three thousand soldiers while Howe wintered with up to ten times that number in and around Manhattan. "These are the times that try men's souls," Thomas Paine wrote during that winter in his pamphlet *The American Crisis*.

WASHINGTON USED THE BLEAK WINTER at Morristown to reassess and revise his army's structure and strategy. Both were faulty. "The misfortune of short Inlistments, and an unhappy dependance upon Militia, have shewn their baneful Influence at every period," he wrote in January 1777, but "at no time, nor upon no occasion were they ever more exemplified than since Christmas. . . . All our movements have been made with inferior numbers, & with a mix'd, motley crew; who were here today, gone tomorrow."[86] The militia, Washington complained, "come in, you can not tell how—go, you cannot tell when—and act, you cannot tell where—

consume your provisions—exhaust your Stores, and leave you at last at the critical moment."[87] He wanted what amounted to a standing army with more soldiers, multiyear terms of enlistment, and severe penalties for disobedience and desertion. For the Continental Army, those extended terms typically became "three years or during the war," which would create confusion when the war unexpectedly lasted more than three additional years.

In addition to re-forming his army, Washington now favored adopting a so-called Fabian military strategy, named after the Roman general Quintus Fabius who wore down a superior Carthaginian army through a war of attrition. "Desperate diseases, require desperate remedies," Washington wrote to Hancock.[88] Congress had opposed a standing army on republican grounds and repeatedly directed Washington to defend major cities rather than fall back to preserve his army. Now, faced with the army's imminent collapse, it agreed to Washington's terms and empowered him to impose them without oversight.[89] "Our Fabius will be slow, but sure," John Adams assured his wife.[90]

By the spring of 1777, Britain had a large army in Canada and a larger one in New York. The United States could only respond with the remnant of the force that once invaded Canada holed up in a defensive posture at Fort Ticonderoga and the remaining troops of the now Fabian-minded Washington encamped in New Jersey. Fully expecting to suppress the rebels by fall, the British began their offensive in June with John Burgoyne's northern army moving south from Canada and retaking Ticonderoga. With the veteran officer Horatio Gates taking command and augmented by state and local militia streaming in from New England and New York plus crack units sent north by Washington, the Americans regrouped in the forested regions around Saratoga. They hoped to stop the British advance before it reached the lower Hudson Valley and split the states in two from Montreal to Manhattan.

Washington and other patriot leaders expected General Howe to send forces north from New York City to relieve Burgoyne. Burgoyne thought so too and did not keep his supply lines open to Canada. Instead, Howe dithered until mid-July trying to lure Washington into an open fight and, when this failed, then loaded two-thirds of his men onto ships headed south. "Howes Fleet has been at Sea, these 8 days. We know not where he is gone," Adams wrote on July 30. "Some guess he is gone to Cheasapeak, to land near Susquehanna and cross over Land to Albany to meet Burgoine. But they might as well imagine them gone round Cape horn into the South Seas to land at California."[91] In fact, failing to recognize that armies (not cities) were critical in this war, Howe had bypassed Washington's forces to assault Philadelphia from the south.

Washington shifted his army to counter Howe's force at Brandywine Creek south of Philadelphia. When this failed, however, he did not follow up with a determined defense of the city. The British occupied it without resistance on September 26. By then, Congress had moved again, this time to central Pennsylvania, where it remained for months. After a sharp clash at Germantown in October, Washington withdrew to winter quarters at Valley Forge while Howe wintered in Philadelphia. Franklin immediately recognized the shortsightedness of Howe's occupying a city with little strategic importance, no established defensive perimeter, and many indifferent Quaker residents. "You mistake the matter," he told friends who lamented the city's capture, "instead of *Howe* taking *Philada.—Philada* has taken *Howe*."[92] Once the operation ended, London recognized the blunder of the British army's not remaining based on Manhattan island and replaced Howe with Henry Clinton, who withdrew his army to New York and later dispatched much of it to the southern states while keeping his headquarters in Manhattan.

Franklin received confirmation that Philadelphia had fallen from the same messenger who brought news that Burgoyne had surrendered with his entire army to Gates at Saratoga. Without relief from the south, Burgoyne's position had become untenable. Suddenly Philadelphia did not matter. With the victory at Saratoga, Franklin had everything he needed to seal an alliance with France. It still required a bit of acting by Franklin, who forced Vergennes's hand by pretending to discuss terms with an anxious British agent, but that ruse only accelerated the inevitable. In Paris on February 6, 1778, representatives of France and the United States signed treaties recognizing American independence, opening formal trade between the two countries, and establishing terms for their military alliance. These treaties did not win the war—that process would take five more years—but they made victory possible. In marked contrast with others at the ceremonial signing, Franklin reportedly sported an old coat for the occasion—the same one that he had worn in 1775 during his denouncement before Britain's Privy Council. He never forgot or forgave that act. The French alliance was his revenge, and he savored it.[93]

WASHINGTON AND FRANKLIN COMMUNICATED by mail more often and more regularly during the year leading up to the French alliance than during any other period, and more letters between the two men survive from the term of Franklin's service in France than from all other times. More than twenty such letters exist from 1777 alone. Of course, Franklin and Washington did not need to communicate by mail when they served together in Congress or, later, at the Constitutional Convention. They could talk. But they did not need to exchange letters during the war years either.

With both men holding positions under Congress during Franklin's tenure in Paris, each typically conveyed critical information to congressional committees or to the president of Congress, who in turn passed it on to the other as needed. Reports, dispatches, and commentary therefore normally passed between the two men through Philadelphia rather than by direct correspondence. As Washington noted in one 1780 letter to Franklin in words similar to what appeared in others, "I doubt not you are so fully informed by Congress of our political and Military State that it would be superflueus to trouble you with any thing relating to either." Nevertheless, he closed this letter with the warm valediction, "With my best wishes for the preservation of your useful life and for every happiness that can attend you which a sincere attachment can dictate."[94]

Franklin replied in kind, with closings such as "My best Wishes always have and always will attend you, being with the greatest and most sincere Esteem & Respect."[95] Rather than terms of endearment, these were words of mutual admiration and reliance.

Although warm words were not uncommon in letters passing between Franklin and Washington during this period, much of their correspondence was purely perfunctory, especially if it related to Europeans seeking commissions in the Continental Army. From his arrival in Paris, Franklin was besieged with supplicants of this sort. Most wanted to be officers and asked him for letters of recommendation to Washington. "These Applications are my perpetual Torment," Franklin complained in 1777. "Not a Day passes in which I have not a Number of soliciting Visits besides Letters."[96] He sought to filter the applicants and to recommend only the best, but some had contacts in France making them hard to refuse.[97] These often received the most cursory letters, including a phrase such as "He goes over at his own Expence, and without any Promise from me."[98] Indeed, Franklin parodied the process in a printed form recommendation. "The Bearer of this who is going to America, presses me to give him a Letter of

Recommendation, tho' I know nothing of him," it stated. "I must refer you to himself for his Character and Merits, with which he is certainly better acquainted than I can possibly be."[99]

Swamped with applicants for commissions from multiple sources, in August 1777, Washington begged Franklin not to send more. This slowed the flow, but some of the most promising candidates still got through. A few whom Franklin recommended with added emphasis proved invaluable: Count Pulaski, Baron von Steuben, and the Marquis de Lafayette. "I give no Expectations to those who apply," Franklin assured Washington. "I promise nothing."[100] This was a working partnership.

Franklin and Washington knew that by being the first, they were establishing norms for future American diplomats and generals, and that these norms called for working through Congress. Certain subjects called for direct communication, however. In one letter from 1780, for example, Franklin invited Washington to tour Europe with him following the war. "You would on this Side the Sea, enjoy the great Reputation you have acquir'd," he wrote in words that surely cheered Washington during a dark period of the war. "At present I enjoy that Pleasure for you: as I frequently hear the old Generals of this martial Country, (who study the Maps of America, and mark upon them all your Operations) speak with sincere Approbation & great Applause of your Conduct, and join in giving you the Character of one of the greatest Captains of the Age."[101] With experience and through mistakes, Washington was learning how to lead soldiers in battle and armies in war—and Franklin wanted him to know it was recognized and appreciated.

Another direct exchange between the two patriot leaders occurred after a massive joint Franco-American operation led to the surrender of Lord Cornwallis and his southern army at Yorktown in October 1781. By fast frigate, Washington immediately sent Franklin a copy of the articles of capitulation and a summary account of the eight thousand prisoners and more than two hundred cannons taken. Transmission "thru the usual channel of the department of foreign affairs" would take too long, Washington explained.[102] With the siege of Yorktown that netted Cornwallis's surrender, the Virginian finally achieved that one bold stroke to sweep Britain from the field that he had yearned for since assuming command of American forces after the Battle of Bunker Hill. But while Washington brilliantly orchestrated the siege, only French troops and ships made it feasible.

Indeed, it is fair to call it a Franco-American operation (with France named first) because more French troops were involved in it than Continental ones. The reference to Washington's yearning for bold strokes after the Battle of Bunker Hill is also apt because, as in the siege of Boston, the forces on the American side at Yorktown outnumbered those on the British side by two to one, with the British trapped on a peninsula. The key difference between the two sieges was that this time, instead of a British navy in Boston harbor able to resupply and ready to evacuate its forces, the British faced a French navy at Yorktown blocking their resupply and escape.

At Yorktown, unlike Boston, Washington could follow his instincts by bombarding and attacking at will to compel a surrender that he reasonably believed would end the war. Fittingly, he shared the news with Franklin. After all, both of them played vital parts in achieving the victory. For his part, Franklin much later wrote to Washington congratulating him "on the final Success of your long & painful Labours in the Service of our Country, which have laid us all under eternal Obligations."[103]

Five

"THE MOST AWFUL CRISIS"

THE LAST FEW PAGES READ as if Franklin's French alliance led directly to Washington's victory at Yorktown and the war's end with virtually nothing of significance in between. Some modern accounts make the story almost this straightforward. Walter Isaacson's masterly biography of Franklin speaks of the alliance helping to "seal the course of the Revolution" and never mentions another battle until Yorktown.[1] (It had named many prior ones.) *The American Revolution: A History,* a primer by the war's preeminent historian, Gordon Wood, races from the alliance to Yorktown in under four pages. The book's telling of the story from Lexington to the alliance spans nearly thirty.

Even at the time, it seemed as if it should be so. The *Pennsylvania Packet,* for example, depicted the alliance as the instrument "of giving liberty, independence, and the prospects of peace to this country."[2] Similarly, the Massachusetts Board of War responded to news from Franklin of the alliance by exalting, "France has seizd the happy moment and perhaps by her new ally may lay a foundation, at least, to prevent Brittain longer ruling the World."[3]

Franklin had all but invited such hopes by his letter to the board, which declared that, under the alliance, the French king "guarantees the Liberty, Sovereignty, and Independence absolute and unlimited of the United States."[4] Such a guarantee, if given by the treaties, would justify great expectations. Washington encouraged them in his general orders to the troops announcing the alliance, which thanked divine providence for "raising us up a powerful Friend among the Princes of the Earth to establish our liberty and Independence upon lasting foundations."[5] Indeed, tying the war's end to the French alliance and victory at Yorktown was implicit in Adams's famous rant, set forth in chapter 1, that Americans would remember their revolution as simply *Dr Franklins electrical Rod, Smote the Earth and out Spring General Washington.*"[6]

Reality was not so simple. More than three and a half years of warfare separated the French alliance in February 1778 from Cornwallis's surrender at Yorktown in October 1781, and nearly another two years passed before the Treaty of Paris formally ended the American Revolution in September 1783. More American soldiers were killed, wounded, and captured after the alliance than before it. Although the winter quarters at Morristown in 1776–1777 and Valley Forge a year later were aptly depicted as times that tried men's souls, the winter that followed in 1779–1780, again at Morristown, was far worse in terms of human privation, despite French aid resulting from the alliance. Bitter cold and heavy snow, including seven blizzards in December and a record thirty-eight snowfalls overall, kept supplies from reaching the encampment and kept soldiers largely confined to some one thousand crudely constructed log huts, each housing twelve men in one small room. "The oldest people now living in this Country do not remember so hard a Winter," Washington wrote near its end. "The severity of the frost exceeded anything of the kind that has ever been experienced in this climate before."[7]

The suffering caused by a lack of food, clothing, and shelter was compounded by the collapse in value of the Continental currency and inability to pay the troops. The ongoing war and Congress's failure to capitalize on the alliance with France to establish a governing structure capable of addressing domestic needs carried essential lessons. Washington, Congress's superintendent of finance Robert Morris, Washington's aide-de-camp Alexander Hamilton, and other budding federalists learned these lessons more during the dark days after the alliance than those before it. And if "federalist" here means those who wanted a stronger federal union, even Franklin fell in this camp. These days shaped their approach to the ensuing peace.

THE CONTINENTAL CONGRESS that appointed Washington as commander in chief and sent Franklin to France had dubious authority to do either (or any) act. Neither the king nor Parliament recognized the Congress, even though it met for more than a year prior to declaring independence. Colonial assemblies chose some of its members (like Franklin), but never with the approval of their appointed governors. Many of its members (like Washington) were chosen by extralegal conventions composed of most (but not all) delegates to a colony's assembly meeting in defiance of their royal governors. Acting under its own authority and by its own rules, the Second Continental Congress managed the war, issued currency, commissioned officers, entered into contracts, borrowed money, negotiated with foreign nations, invaded Canada, and declared independence. So much for divine right of kings; here was republicanism with a vengeance.

Anticipating challenges to Congress's legal authority, in 1775, Franklin introduced in Congress an enhanced version of his 1754 Albany Plan of Union. This would have provided a governing authority for Congress, but it never received a hearing. Only after Congress began debating independence in June 1776 did it appoint a committee to draft articles of confederation. Committed to a stronger central government than the committee proposed, Franklin was deeply involved in congressional debates over those articles until Congress dispatched him to France in September 1776.

After deliberating off and on for more than a year, Congress approved the Articles of Confederation in November 1777 and sent them to the states for ratification. Designating the union as "a league of friendship" among sovereign states, the Articles formed a weaker central government than either Franklin or Washington wanted.[8] It lacked power to levy taxes, draft troops, or pass laws for the general welfare. The states (rather than the people) were represented in Congress, with each state having one vote, a supermajority of nine votes required to pass legislation, and unanimity needed to amend the Articles. Franklin had favored proportional representation so that people (rather than states) would be represented. The Articles limited Congress's authority to matters of war and peace, foreign affairs, relations with Native peoples, postal services, standardizing weights and measures, and resolving disputes between states. Even then, given the reluctance of states to relinquish power, final adoption took time. Ten states ratified within five months but concerns about the status of western lands kept the final state, New Jersey, from agreeing until 1781. In the meantime, Congress followed the Articles as the de facto frame of government. Congress's weakness frustrated both Franklin and Washington. It made their jobs more difficult.

WASHINGTON FELT IT MOST. The French alliance opened all sorts of opportunities for his army, but a chronic lack of resources hobbled its ability to exploit them. Charged following the alliance with also defending Britain's prized sugar-producing colonies in the West Indies against French assaults and Florida from recapture by France's ally Spain, British commander Henry Clinton withdrew his troops from Philadelphia in 1778 to consolidate them with those remaining in New York before dispatching many of them to the south. Buoyed by the alliance and fresh from a winter of drilling the troops at Valley Forge in the ways of European warfare under the tutelage of Baron von Steuben, Washington opted to harass the British on their way to New York. One of the European officers recommended by Franklin to Washington, von Steuben had been hounded out of the Prussian army by charges of homosexuality and arrived with a young male companion at Valley Forge, where no questions were asked.[9] The winter training proved its worth in New Jersey in June 1778, when pursuing troops under Washington held their ground against a British counterattack at the Battle of Monmouth, before the enemy disengaged after nightfall and hurried on to New York.

For the rest of the war, Monmouth was the closest that Washington came to winning a battle against the British in the north. His army was chronically undermanned, poorly clothed, short on equipment and supplies, and sometimes near starvation. Soldiers and officers alike often went without pay or were paid in nearly worthless Continental script. If an army travels on its stomach, as Frederick the Great of Prussia allegedly said at roughly this time, then it was amazing the Continental Army covered any ground at all. During that awful second winter at Morristown in 1779–1780, Washington imposed forced requisitions of livestock and grain on local farmers and temporarily turned a blind eye to plundering by starving soldiers. They had survived on half rations since fall but by winter, the army's exhausted stores could not supply even this reduced amount. With heavy snows blocking access, hunger thinned the troops. By spring, desertion was rampant and mutiny in the air.

The two summers between the Battle of Monmouth and that second winter at Morristown saw nothing but stalemate or setback for Washington's army in the north, despite the redeployment of more than half of the British army southward. Once Clinton had concentrated his remaining troops in New York, Washington spread his army in a wide arc around the city from western Connecticut to northern New Jersey. He dreamed of the French navy blocking the harbor and his men storming Manhattan, but Washington never had enough strength to conduct anything more than borderland skirmishes in the war-ravaged lower Hudson River valley. The sole joint Franco-American operation in the north during this period, an assault on British-held Newport in 1778, bogged down after a gale dispersed a French squadron before it could engage the British navy off the Rhode Island coast. Yet a year later, the British abandoned Newport to send more forces south. With both sides adopting something of a Fabian strategy in the north, 1780 came and went with the two sides roughly where they had been following the Battle of Monmouth some two years earlier, and a sizable portion of Washington's army settling in for a third hard winter near Morristown. After visiting those troops, Washington established his own winter headquarters at New Windsor in the lower Hudson River valley.

Although not as harsh as the prior one, the 1780–1781 winter encampment near Morristown came close to breaking the army. It followed another dispiriting year. Going on the offensive in the south, British forces captured Charleston and much of South Carolina during the spring of 1780. An American army under Horatio Gates assigned to regain South Carolina was routed at the Battle of Camden in August 1780, when militia on the left flank fled without a fight. Their flight further convinced Washington that only disciplined Continental soldiers led by an elite officer corps could stand against British regulars.

Then, only a month later, disillusioned in part by his own and the army's treatment by Congress, one of Washington's most trusted officers, General Benedict Arnold, offered to sell the center of the American defensive line around New York at West Point for £20,000 and a commission in the Royal Army. Exposed by chance on the eve of its consummation, giving Arnold just enough time to escape, the plot shocked soldiers and officers alike. Not only had Arnold been a hero of Ticonderoga and Saratoga, he was gravely wounded in the assault on Quebec after leading a death-defying late fall march to it.

"Treason of the blackest dye was yesterday discovered!" Washington wrote in his general orders for September 26, 1780. "General Arnold who commanded at Westpoint, lost to every sentiment of honor—of public and private obligation—was about to deliver up that important Post into the hands of the enemy. Such an event must have given the American cause a deadly wound if not a fatal stab."[10]

When he heard about it in faraway Paris, Franklin said of Arnold, "His Character is in the Sight of all Europe already on the Gibbet & will hang there in Chains for Ages."[11]

Characteristically, Washington attributed exposure of the plot to providence—"convincing proof that the Liberties of America are the object of divine Protection," he wrote—but could not as easily blame the betrayal on demonic forces.[12] The despair reflected in Arnold's act had human causes that went beyond a single man. Even Washington felt it.

CONGRESS HAD FAILED THE TROOPS. Food was scarce, clothing scarcer, and pay scarcest of all. The Continental currency had become a running joke, war profiteers reaped large returns, and the army limped along at less than half its authorized size.

At first Washington blamed the low caliber of the members serving in Congress. Good people had served at the outset, he knew—people much like himself and Franklin. Their number included John Hancock, John Adams, Roger Sherman, George Wythe, and Thomas Jefferson. By the end of 1778, however, Washington was asking, "Where is Mason—Wythe—Jefferson?" They were serving in state government, not Congress. "I think our political system may, be compared to the mechanism of a Clock," he added in a telling Enlightenment era metaphor, "and . . . it answers no good purpose to keep the smaller Wheels in order if the greater one which is the support & prime mover of the whole is neglected." Given one wish for America, he wrote, "I shall offer it as mine that each State wd not only choose, but absolutely compel their ablest Men to attend Congress." None did, and the army suffered from galloping inflation and a lack of resources. "A great part of the Officers of yr army from absolute necessity are quitting the Service and the more virtuous few rather than do this are sinking by sure degrees into beggery & want," he warned a former congressman.[13]

In a development carrying profound implications for America's political future, Washington soon realized that the system was more to blame than individuals. The Congress simply lacked sufficient power to execute the war or serve the common cause. "In modern wars the longest purse must chiefly determine the event," Washington said of Britain. "Their system of public credit is such that it is capable of greater exertions than that of any other nation."[14] In stark contrast, Congress relied on requisitions to the states for money and troops, without power to enforce them. "One state will comply with a requisition of Congress—another neglects to do it. a third executes it by halves—and all differ in the manner," Washington complained. "I see one head gradually changing into thirteen."[15] In a July 1780 letter pleading for the sort of federal union later forged by the Constitution, he argued that unless Congress obtained "absolute powers in all matters relative to the great purposes of war, and of general concern (by which the

States unitedly are affected . . .) we are attempting an impossibility" in seeking and sustaining independence.[16] "Our Cause is lost" without an effective union, he declared.[17] "The contest among the different States *now,* is not which shall do most for the common cause, but which shall do least."[18]

The reactions of others in the Continental Army during these years of military stalemate and congressional inaction typically fell somewhere between the perfidy of Arnold and the protests of Washington. A few, like Washington, learned lessons in the importance of an effective central government with a strong system of public credit: Hamilton leaps to mind. Many, also like Washington, grumbled against public ingratitude and private profiteering.[19] And some, unlike Washington, deserted or quit when their terms of enlistment ended—leading to something of a revolving door for service and enhanced inducements for recruiting soldiers and retaining officers. In 1778, for example, Congress acceded to the demands of officers (backed by Washington) that, for continued service to the war's end, they (like British officers) would receive half-pay pensions. As the war wore on, many states offered hefty bounties for enlistment and began drafting soldiers to fulfill manpower requisitions for the Continental Army. The draft laws, like those from colonial days, allowed draftees to hire a replacement—a feature even Washington was willing to use to keep his nephew at Mount Vernon managing the plantation. As a result, by 1780, the Continental Army looked more like its professional British counterpart, with an elite officer corps and hardscrabble soldiery, than the republican citizens' militia of 1776. Perhaps the new army fought for liberty, but it also expected pay.

These developments form the backdrop of the dark winter of 1780–1781 in Morristown. Three years had passed since soldiers there from Pennsylvania had enlisted for "three years or during the war." With their state now offering added rewards to new recruits, relying on the "three year" term in their enlistment contracts, these veterans demanded similar bounties to remain in service. Desperate for soldiers and devoid of resources, Washington relied on the "or during the war" wording to order continued service. Mutinying, the armed men marched on Philadelphia. Literally and figuratively, they were met halfway, in Trenton, by the president of Pennsylvania offering them release from service or added compensation, with most choosing the payment. When New Jersey troops in Morristown threatened a similar uprising, Washington had two ringleaders summarily shot. Without swift punishment, he warned Congress, "The infection will no doubt shortly pervade the whole mass" bringing "an end to all subordination in the Army, and indeed to the Army itself."[20] By soldiering on for one more year, Washington's army, destitute and half naked, turned the world upside down.

BELEAGUERED DUE TO A LACK of fiscal accountability and political responsibility by the same enfeebled Congress that bedeviled Washington, Franklin did as much as anyone to bring about the momentous events of 1781. Soldiering on for three years after signing the treaty with France in 1778, he held the alliance together and served as America's senior diplomat in Europe, despite his advanced age and crippling bouts with gout and gall- or kidney stones. Once France recognized American independence, Congress named Franklin as the country's sole minister to the court of Louis XVI. The prior triumvirate had proved awkward anyway, with each commissioner having his own ideas but only the celebrated Franklin having Vergennes's ear. The embittered Francophobe Charles Lee had assailed Franklin as a doddering dupe of France. Matters had only grown worse when the proud and puritanical John Adams replaced Silas Deane as the commission's third member and began questioning Franklin's fraternizing with the French—"a Scene of continued discipation," Adams said[21]—as well as his fidelity to America. Franklin had suffered it all with remarkable equanimity while performing his diplomatic duties (which he knew involved wining, dining, and flirting) with extraordinary success. He pitied the small-minded Lee and dismissed Adams as "sometimes and in some things, absolutely out of his Senses."[22]

During the years between Saratoga and Yorktown, Franklin had to maintain the alliance while continually asking the French for financial aid, naval support, arms, equipment, and troops. In addition, he arranged shipments of uniforms for the poorly clothed and partly shoeless American army, purchased war matériel from European suppliers, and negotiated for the care and release of prisoners of war held in Britain. Franklin also oversaw American navy ships and privateers operating from France and aided American merchants doing business in Europe. He vetted European military officers seeking commissions in the American army, helped American states secure loans from France, and conducted back-channel peace talks with British contacts. He even printed American passports and other documents on his own printing press. Not counting his active social life, which helped the cause by winning friends in high places and securing information, Franklin reported that he never worked harder in his life. "Certainly no one else could have represented America abroad as Franklin did," historian Gordon Wood concluded. "He was the greatest diplomat that America has ever had."[23] The fact that the American army survived until 1781 despite its lack of food, clothes, and pay was partly due to Franklin. The promise that a French force would join it for the siege of Yorktown was largely due to him.

Most amazing of all, Franklin achieved all this without having an effective nation to represent. Like Washington, in his official capacity when speaking of the collective, Franklin used the term "Americans," not "Virginians" or "Pennsylvanians." He envisioned the United States as a national union of people, not a confederation of states—or at least that is how he wanted to view it. At the outset of the Revolutionary War, by asserting the power to form an army, appoint generals, print money, demand and collect funds from the states, name diplomats, and declare independence, the Continental Congress acted as if it led a nation. Able leaders from the various states vied to serve in Congress and, if chosen, actively participated in its affairs: Washington, Franklin, Hancock, Morris, the Lees of Virginia, the Adamses of Massachusetts, Dickinson, Jefferson, and more.

As time wore on and the Articles of Confederation began setting the tone, power shifted to the states, which could levy taxes and conscript soldiers. The caliber of the delegates in Congress, the influence they carried in their states, and the level of their contributions steadily waned. So many members did not even show up that, from 1777 on, Congress never had all thirteen states represented at any one time and sometimes lacked a quorum of seven. Some states, like South Carolina, started their own navies and had the audacity to ask Franklin, who was trying to secure warships for Congress, to help procure them for state fleets. And money, always money. Congress needed loans from France because it could not levy taxes and, by 1779, its paper money was worthless, but the states wanted loans too. "The Agents from our different States running all over Europe begging to borrow Money," Franklin complained, "has excedingly hurt the general Credit, and made the Loan for the United States almost impracticable."[24] Much as Washington—the leader of its sole effective instrumentality, the army—came to symbolize the United States at home, Franklin did so abroad. Only he could have obtained continued credit for a disintegrating union.

COMING OUT OF THE GRIM WINTER OF 1780–1781, prospects for American independence looked worse than ever. On the heels of the short-lived mutinies by troops from Pennsylvania and New Jersey, crippling desertions, raging camp illness, and the lack of pay for officers and men, Washington wrote to his former aide-de-camp John Laurens in April 1781 "that without a foreign loan our present force (which is but the remnant of an Army) cannot be kept together this Campaign; much less will it be encreased, & in readiness for another."[25] The army stood at about one-third of its authorized size. Many of the soldiers lacked a shirt, or shoes, or both.

Congress had dispatched the twenty-six-year-old Laurens, the wealthy and winsome son of its former president Henry Laurens, to Paris to help Franklin secure an added loan from France by relating dire conditions in the field. With Congress utterly bankrupt by 1781, its leaders did not believe they could carry on the war without an immediate infusion of cash and feared that the aged Franklin could no longer secure the funds.[26] Laurens did have a boyish enthusiasm that had charmed Washington and led to a close relationship with Hamilton in camp, but he lacked any diplomatic skills whatsoever and, with his impetuous demands, quickly alienated Vergennes.

Fortunately, Franklin had secured a gift of six million *livres* from Louis XVI literally days before Laurens arrived. In the letter reporting this remarkable achievement to Congress, Franklin turned the tables on his critics by offering his resignation. At age seventy-five and after fifty years of public service, "I do not know that my mental Faculties are impair'd; [but] perhaps I should be the last to discover that," Franklin wrote in an indirect rebuke of anyone who questioned his loyalty or ability.[27] "I fancy it may have been a double Mortification to those Enemies," he later confided in a friend, "that I should ask as a Favour what they hop'd to vex me by taking from me." Of course, Congress insisted that he continue in office. "I call this Continuance an Honour, & I really esteem it to be a greater [one] than my first Appointment," Franklin wrote with some satisfaction.[28]

FRANCE'S BENEFICENCE COINCIDED with an unexpected upturn in American fortunes. After taking Charleston in 1780, the British general Henry Clinton left the task of rolling up the southern colonies to the brash aristocrat Charles Cornwallis, who took apparent pleasure at punishing rebels in what devolved into near guerrilla warfare for the rural Carolinas. He lost a supporting loyalist militia force at the Battle of Kings Mountain that October and a detachment of regulars at the Battle of Cowpens early in 1781 before suffering heavy casualties in a Pyrrhic victory over a combined force of continentals and state militia under General Nathanael Greene at the Battle of Guilford Courthouse in March.

Unbowed by these setbacks—indeed, unwilling to see them as serious setbacks—Cornwallis marched his army north to Virginia, where he joined with a small British force under Benedict Arnold, which was already wreaking havoc in the state. Hoping to crush the Revolution in its Virginia heartland, Cornwallis fell back in August to the coast at Yorktown, on a tidewater peninsula in the Chesapeake Bay, where he confidently assumed that the Royal Navy could resupply and, if needed, evacuate his army.

Responding to Franklin's pleas, France had dispatched an army commanded by Comte de Rochambeau to Rhode Island in 1780 and a fleet under Comte de Grasse to Virginia in August 1781. Washington initially hoped to combine his army with Rochambeau's to liberate New York but could not succeed without sea power. Upon learning of de Grasse's plans, he shifted all available French and American troops to Virginia with the aim of trapping Cornwallis's army at Yorktown.

After the French navy blocked any escape or resupply by sea, some sixteen thousand French and American soldiers surrounded about eight thousand British troops at Yorktown and commenced classic European siege operations in September. Franklin received word of these developments from Congress's new foreign secretary, Robert Livingston, in a message sent just before the trap sealed shut. "The enimy have evacuated their principal outworks," Livingston wrote in mid-October, "and the least sanguine among [our] Officers fix the end of the month, as the era of Cornwallis's captivity."[29] Upon receiving this letter, Franklin promptly alerted Vergennes, who, on the same day and with later news, informed Franklin of the British capitulation. A messenger sent by Washington soon confirmed the report. With his army's outer defenses overrun, trenches reaching ever nearer the inner lines, and cannons shelling Yorktown from close range, Cornwallis had surrendered his entire force on October 19, 1781. Barred by the terms of surrender from displaying colors or performing martial music, legend has the British marching out of Yorktown with their bands playing the tune to a popular song, "The World Turned Upside Down." To them, it was. "The Success has made Millions happy," Franklin reveled in his reply to Vergennes.[30]

FOLLOWING THE SIEGE OF YORKTOWN, letters of congratulation flowed to both Washington and Franklin. Those to Franklin from such New Englanders as Massachusetts governor John Hancock urged him to secure fishing rights in the Grand Banks off Newfoundland as part of any peace treaty with Britain. These correspondents simply assumed that Franklin would quickly negotiate the war's end once Cornwallis's army had surrendered.

Franklin did not have that luxury. While Yorktown sapped the British public's will to fight, it did not break the king's resolve. British troops still held entrenched positions in New York, Charleston, and Savannah that the Americans could not dislodge, and the Royal Navy continued to fight France and its allies, Spain and the Netherlands, in the Caribbean and elsewhere. So long as he retained a supportive government in Parliament under Lord North, George III refused to give up. Worried about his reputation, he confessed to praying "that posterity may not lay the downfall of this once *respectable* empire at my door."[31]

Complicating matters, in 1779, Congress had entrusted peace negotiations to the vain and distrustful John Adams, who soon antagonized both the British and French. They preferred dealing with Franklin. Vergennes pressed this point on Congress, which in June 1781 expanded its peace commission to include Franklin, Jefferson, Henry Laurens, and the American minister to Spain, John Jay, as well as Adams. By this point, Adams had moved to the Netherlands in order to seek loans for Congress from Dutch bankers, who found him as difficult to deal with as the French and British had. When North's government finally fell in March 1782 and peace negotiations began in earnest, at least Adams's mission to Holland kept him otherwise occupied.

Aside from directing their commissioners to follow France's lead, Congress gave them a free hand in negotiating a peace treaty with Britain. Since the negotiations occurred in Paris, this left preliminary matters to Franklin. Jefferson never left Virginia. Having been captured by the British during a transatlantic crossing in 1780, Laurens was still imprisoned in the Tower of London at the time of his appointment and did not reach Paris until late in the negotiating process. Diplomatic duties kept Jay in Spain and Adams in Holland well into 1782, with Jay incapacitated by illness when he did arrive. By that time, Franklin had struck an initial understanding with British negotiators. At first, he had deferred to France even as Britain sought a separate peace, but after Vergennes became impatient to end the war and authorized him in May to negotiate independently to speed the process,[32] Franklin pressed ahead with remarkable success.

Franklin was a straightforward negotiator who stated his terms and held to them. He set forth four necessary conditions for any treaty: full and complete independence for the United States, removal of all British troops from the states, fishing rights for Americans on the Grand Banks, and recognition of borders prior to the Quebec Act of 1764. This last term was critical because (even though American forces had done little to win the west) it would extend the states to the Mississippi River rather than cut them off at the Appalachian Mountains. Like Washington, Franklin believed in the frontier. As a young man, he had gone west, to Pennsylvania, to make his future. Before inheriting Mount Vernon, Washington sought his fortune as a frontier surveyor. Both had invested in western lands and viewed the frontier as a source of economic opportunity and republican virtue. In support of his fourth demand, Franklin warned British negotiators that Americans were restless people destined to push westward. If Britain retained the Ohio Country, tensions would inevitably arise on the border and sour relations between the two nations. Franklin also suggested some advisable terms for the treaty, such as an apology and reparations from Britain, a free-trade agreement between the countries, and the cession of Canada, as ways for Britain to regain the affections of Americans. Predictably, Britain agreed only to his necessary conditions but did concede to all of those.

A final treaty needed to wait for France and its allies to reach their own agreements with Britain. Spain wanted Gibraltar but settled for reclaiming Florida. The Netherlands regained territory in the East Indies. France, which bore the brunt of the war costs, ultimately got the least—only the West Indian island of Tobago and the return of a trading post in Senegal.

Franklin had cut the best deal of all for the United States—perhaps the finest in the history of American diplomacy. When Jay and Adams arrived, they tried to sweeten the terms in further negotiations, but achieved little. In November 1782, they joined Franklin in signing preliminary articles of peace with British negotiators along the lines previously struck by Franklin. Upon learning that the Americans had signed a preliminary peace agreement, Vergennes voiced displeasure at not being advised in advance. Franklin managed to mollify him to such an extent that France extended Congress yet another loan to tide it over until the Peace of Paris officially ended the war in September 1783. Until then, the British continued to disrupt American shipping on the high seas and occupy New York City, Charleston, and Savannah. Washington's army had in the meantime settled back into its positions in the lower Hudson Valley around Newburgh to guard against any outbreak by the British in New York.

CONGRESS NEEDED THE ADDITIONAL LOAN because, rather than strengthening its hand, the diminishing military threat from Britain and impending peace made the states even less willing to contribute to the collective cause than before. Having gone with part or no pay for years, Washington's soldiers worried that they would never get paid. His officers feared not only for their back pay but also for their hard-won pensions. Once Franklin and his fellow peace commissioners secured a final treaty with Britain, Congress could afford to ignore veterans and devote its scarce resources to future needs or its most powerful creditors. The resulting tinderbox of armed, idle troops bearing legitimate grievance against Congress severely tested republican norms of civilian control over the military that Washington had vowed to respect. Individual units had mutinied for pay before but now, at Newburgh, Washington feared an army-wide uprising.

"When I see such a number of Men goaded by a thousand stings of reflexion on the past, & of anticipation on the future, about to be turned into the World, soured by penury & what they call the ingratitude of the Public, involved in debts, without one farthing of Money to carry them home, after having spent the flower of their days & many of them their patrimonies in establishing the freedom & Independence of their Country," Washington warned Congress in October 1782, "I cannot avoid apprehending that a train of Evils will follow."[33] Without knowing specifics of the situation except through grim reports from Henry Laurens, Franklin also worried about the soldiers' plight and its possible consequences. It spurred his pleas for French aid at a time when the financial exigency at Louis XVI's court was such that even payment to French officers in America was delayed by a year.

With Washington's tacit approval, during the closing days of 1782, a delegation of officers from Newburgh carried a petition to Congress, which then met in Philadelphia. The petition appealed for the ascertainment of the amount owed each officer for back pay and expenses, with security established for future payment. "We have borne all that men can bear—our property is expended—our private resources are at an end," the officers grumbled in the petition. "The uneasiness of the soldiers, for want of pay, is great and dangerous; any further experiments on their patience may have fatal effects."[34]

Upon the delegation's arrival, the soldiers were embraced by Robert Morris, Congress's superintendent of finance and a leading advocate of a strong central government. Adding to the warmth of its welcome by federalists like Morris, the delegation reached Philadelphia only days after Congress learned that the states had failed to ratify its proposal for a national tax or "impost" on all goods coming from overseas, which Morris had pushed through Congress as the means to pay past debts and finance ongoing operations. Federalists in Congress saw the officers' petition as a timely tool to revive the impost. All they needed to gain its ratification, some thought, was for the army to link its worthy cause and veiled threats with the political clout of wealthy domestic creditors.[35] More cynical federalists, like Morris's rakish young assistant, Gouverneur Morris (to whom he was not related), privately conceded that it might take an actual show of force by the unpaid troops to wrest taxing authority for Congress from the states. On New Year's Day 1783, the younger Morris wrote to the like-minded John Jay in Paris about soldiers "with swords in their hands" securing that power for Congress "without which the government is but a name," that is, the power to tax.[36]

HISTORIANS HAVE LONG DEBATED exactly what situation Morris had in mind—mere threats or actual insurrection—and how much the two Morrises and Washington's former aide, Congressman Alexander Hamilton, tried to hasten it. Certainly they were wily politicians who fought fiercely for their ends and, at the time, viewed the taxing power as an essential end for a stable government. After conferring with them, some of the officers began warning Congress that the troops might mutiny without pay.

Despite these provocations, Newburgh remained quiet until mid-February, when word reached America that the British government had agreed to independence. Hamilton now wrote to Washington warning him of rumors that some unpaid officers at

Newburgh might reject his republican leadership and use the army to secure their pay and pension by force.[37] Washington could read the ambiguously worded letter as urging him either to channel the insurrection in ways that would strengthen the central government or to squelch it in advance.[38]

Responding to Hamilton, Washington blamed the dissension in camp on General Horatio Gates (though "I have no proof of it," he conceded) and junior officers at Gates's headquarters near Newburgh. Despite this mutinous faction and the injustices suffered by all, Washington assured Hamilton that "the sensible, and discerning part of the Army" would remain subordinate "notwithstanding the prevailing sentiment in the Army is, that the prospect of compensation for past Services will terminate with the War."[39]

Despite these confident words from Washington, rumblings among junior officers at Gates's headquarters threatened to erupt into a general revolt against Washington's leadership and civilian rule in mid-March, following word of a preliminary peace treaty with Britain. The cabal may have included Gates and the Morrises.[40] Hamilton almost certainly knew about it; some historians surmise that he orchestrated it.[41]

On March 10, the conspirators at Newburgh distributed an anonymous call for a meeting of field officers and company representatives on the following day, coupled with an unsigned address outlining their demands. "Suspect the man who would advise to more moderation and longer forbearance"—presumably Washington—and present members of Congress with a stark "alternative," the address demanded. If peace comes without pay, tell them that "nothing shall separate them from your arms but death"; if war continues, "retire to some unsettled country, smile in your turn, and 'mock when their fear cometh on.'"[42]

Washington reacted swiftly. He issued general orders disallowing the anonymously called meeting. Perhaps fearing that it might proceed anyway without an orderly alternative, he authorized a similar one for the fifteenth—the Ides of March—and directed Gates to preside.[43] Officers from every unit in the Newburgh encampment attended that meeting. As soon as Gates called the session to order, Washington entered the hall and asked to speak first.

As was his custom on formal occasions, Washington read a prepared statement. Less than two thousand words long, it spoke of his sacrifice, a soldier's duty, and the impracticability of the conspirators' scheme. To secure the officers' back pay and future pensions, Washington vowed to do so much "as may be done consistently with the great duty I owe to my Country."[44] Given his influence, this promise meant much to his men. It gave them hope for some redress.

After concluding his short but stern speech, Washington reached into his coat pocket for a letter from a friendly congressman, Virginia's Joseph Jones. It reiterated the extreme gravity of the government's financial situation and summarized the ongoing efforts of Congress to address it. Struggling with the handwriting as he read from the letter, Washington drew reading glasses from his waistcoat and asked, "Gentlemen, you will permit me to put on my spectacles, for I have not only grown gray but almost blind in the service of my country."[45] Few of the officers had seen Washington wear glasses. Coming on the heels of his hard address, this soft show of familiarity in their presence moved many. It both humanized and elevated him. Whether from his lofty words or his lowly gesture, some officers wept openly. With this finely timed performance that some historians suspect was rehearsed, Washington carried the day.

After Washington withdrew, the officers present approved resolutions asking him to represent their interests before Congress and rejecting "with disdain the infamous propositions contained in a late anonymous address to the officers of the army."[46] Fifty years later, testifying to its significance in their eyes, survivors of the episode joined a later generation of Americans in erecting on the site a stone obelisk bearing the words "Birthplace of the Republic" for here, they believed, America's republican ideals had been established.

TAKING HIS VOW to champion the officers' cause to heart, Washington began using his platform as America's leading citizen to call for quickly and fairly compensating the troops, and ultimately for building a strong central government that could support those payments and a permanent military establishment. He argued his case privately in a series of letters to members of Congress before taking it public after the states rejected a second, scaled-back import-tax measure. Franklin knew nothing of these events until long after—indeed, he complained in an April 15 letter to Livingston that he had received no news from America for nearly three months—yet he appreciated the gravity of the situation and pressed the case in France for relief.[47]

By early April, Washington became so worried about unrest within the ranks that he counseled Congress to disband the army as soon as possible.[48] Following the official cessation of hostilities between the United States and Britain in mid-April, and with Congress lacking funds to pay the army, Washington endorsed a plan to release the troops with orders that their states (rather than Congress) pay them. Although British forces still occupied New York pending a final peace treaty, Congress responded by furloughing most of the army in June without any cash payment to the departing men. Washington found the episode shameful but necessary.[49]

With the end of warfare and resulting furloughs, Washington had time to focus on the country's future as he waited with a rump force for the final peace treaty and the British to evacuate New York. The process dragged on until late November. During this period, Washington issued the two most significant documents of his military career. While both took years to bear fruit, they helped to lay a foundation for the new constitutional order that Washington would eventually lead.

The first, "Sentiments on a Peace Establishment," argued for a peacetime military of a size and with a mission that would require a robust central government with taxing authority. Such a force ran contrary to conventional thinking about republican rule. "Altho' a *large* standing Army in time of Peace hath ever been considered dangerous to the liberties of a Country," Washington now argued, "yet a few Troops, under certain circumstances, are not only safe, but indispensably necessary."[50] These would be Continental troops organized into regiments to secure the western frontier and guard the border with British Canada. Existing state militias would be restructured on a Swiss model as a uniform force of citizen-soldiers ready for call-up in times of need. Washington also recommended that the central government build a navy to protect merchant shipping, maintain a base at West Point to guard against invasion from Canada, and open an academy to train officers. But his main concern was securing the west, which (like Franklin) he viewed as essential for the country's future. Washington called for a series of army posts along the Ohio River, north to the Great Lakes, and west to the Mississippi. If accepted, his proposal would go a long way toward forging a federal union and giving it a national purpose founded on secure borders, global trade, westward expansion, and imperial prospects.

Washington followed this proposal, which he submitted to Congress, with a so-called circular letter simultaneously sent to all the states urging them to create "an indissoluble Union of the

States under one Federal Head." He explained, "It is indispensible to the happiness of the individual States, that there should lodged somewhere, a Supreme Power to regulate and govern the general concerns of the Confederated Republic." These concerns included military and diplomatic affairs, taxing authority for federal services, and "compleat justice to all the Public Creditors," including unpaid troops. America's independence, as declared in 1776 and later recognized by other nations, rested on a union of the states, Washington observed, and could survive in a hostile world only with union. "The foundation of our Empire was not laid in the gloomy age of Ignorance and Superstition, but at an Epocha when the rights of mankind were better understood and more clearly defined, than at any former period," he added. With reason and experience, Americans could forge a more perfect union—an empire of states capable of taking its place among the great nations of the world.[51] He viewed the United States in that way, regularly calling it an "empire" (rather than a "league of states") during the confederation period.[52] The term was singular, not plural.

This circular letter, which Washington depicted as "the Legacy of One, who has ardently wished, on all occasions, to be useful to his Country," read as his farewell to the people. For the first time in a public statement, he declared his firm intent to retire at war's end. Not after power for himself, Washington noted, "I could have no sinister views" in promoting a strong central government.[53] Hailed as "Washington's Legacy," the letter appeared in newspapers from New England to the deep south and became one of the most celebrated documents of the day.[54] Although none of its nation-building recommendations bore fruit for more than five years, they at once became associated with Washington and linked him in the public mind with the cause of nationhood.

On November 25, 1783, five months after Washington issued this circular letter and eleven weeks after Franklin and others signed the Treaty of Paris ending the Revolutionary War, the British army finally left New York and a residual force of Continental troops bolstered by the New York militia entered war-ravaged Manhattan to the cheers of its long-suffering citizens. During this period, while he spoke openly only about strengthening the Articles of Confederation, Washington privately was calling for "a Convention of the People" to draft a new "Federal Constitution."[55] To survive, he wrote to Lafayette, the United States must "form a Constitution that will give consistency, stability & dignity to the Union; and sufficient powers to the great Council of the Nation for general purposes."[56] To his brother John, Washington added, "Competent Powers for all *general* purposes shoud *be vested* in the Sovereignty of the United States, or Anarchy & Confusion will soon succeed."[57] In his farewell address to the army, issued upon the formal discharge of furloughed troops in early November, he warned, "Unless the principles of the Federal Government were properly supported, and the Powers of the Union increased, the honor, dignity, and justice of the Nation would be lost for ever." Washington urged the departing men—"one patriotic band of Brothers," he called them—to return home as champions of "the Union."[58] He did so himself.

Following the liberation of New York, where he held an emotional final banquet with his remaining officers, and a sentimental journey through battlefields in the middle states, Washington presented himself to Congress on December 23. Only twenty delegates representing but seven states remained in attendance at that little-respected and largely ineffectual body, which was then meeting in Annapolis, Maryland, after protests by unpaid troops drove it from Philadelphia. Even Hamilton and the like-minded federalist James Madison of Virginia had gone home in despair. Addressing its members, Washington treated Congress to the respect that he wanted it to have. "Having now finished the work assigned to me," he stated, "I here offer my commission and take my leave of all the enjoyments of public life."[59] Then he was gone, racing toward Mount Vernon in time to reach it by Christmas. Unlike countless revolutionary generals before and after him, Washington retired.

FRANKLIN'S RETIREMENT TOOK longer to effectuate than Washington's but was similarly final. Citing age, he had offered to retire in 1781, but Congress urged him to stay through the negotiation of peace. "It is the desire of Congress to avail themselves of your abilities and experience at the approaching negotiation," the president of Congress wrote at the time. "Should you find repose necessary after rendering the United States this further service Congress in consideration of your age and bodily infirmities will be disposed to gratify your inclination."[60] Upon sending Congress the signed preliminary peace treaty with Britain in November 1782, Franklin renewed his request to retire, and this time he meant it. Approaching the age of seventy-seven and increasingly incapacitated, he now wanted repose. "If I live to see this Peace concluded, I beg leave to remind the Congress of their Promise then to dismiss me. I shall be happy to sing with Old Simeon, *Now lettest thou thy Servant Depart in Peace, for my Eyes have seen thy Salvation,*" Franklin wrote to Congress's secretary of foreign affairs, quoting scripture for emphasis.[61] At the time, he did not know whether he would return to Philadelphia or remain in Paris. His health might preclude another ocean voyage and he was beloved in France. After more than six years away, Franklin could not know what might await him in America.

Part of his concern about America stemmed from his despair over the confederation's weakness. Even before the Revolution, Franklin had championed a strong central government with control over western expansion and military affairs, power to make laws and tax individuals, and popular representation in a congress. His 1754 Albany Plan of Union, offered during the French and Indian War, contained all of these features. In 1775, Franklin introduced an enhanced version of it as a plan of union to the Continental Congress and, when that body took up consideration of much weaker articles of confederation, he pushed to make them stronger until he left for France. There he suffered under an enfeebled Congress that he was charged with representing before the monarchies of Europe. Franklin was annoyed by the states interfering with his efforts on behalf of the union and pushing their own independent agendas in Europe, frustrated by their independent military programs, and driven to distraction by their unwillingness to pay their requisitions to Congress or give it the power to tax. Even though he served his state in nearly every conceivable capacity, Franklin always viewed himself more as an American than as a Pennsylvanian—and he wondered why others could not see their rational self-interest in a similar way. "No one else except George Washington had had such direct personal experience of the existing government's incapacity," Franklin biographer Edmund Morgan noted.[62]

The inability of Congress to collect taxes upset Franklin most of all. Like Washington, he did not want French troops to fight American battles or French loans to fund the American government—he did not even want an entangling foreign alliance—but (like Washington) he was forced to accept them. In 1782, when (in pleading for yet another loan from France) Robert Morris informed him that the states had paid only $125,000 toward their $8,000,000 requisition for the year (or 1.6 percent), Franklin could not control himself.[63] "I see in some Resolutions of Town-Meetings, a Remonstrance against giving Congress a Power to take as they call it, *the People's Money* out of their Pockets tho' only to pay the Interest and Principal of Debts duly contracted. They seem to mistake the Point. Money justly due from the People is their Creditors' Money," he wrote to Morris. "Property that is necessary to a Man for the Conservation of the Individual & the Propagation of the Species, is his natural Right which none can justly deprive him of: But all Property superfluous to such purposes is the Property of the Publick, who by their Laws have created it, and who may therefore by other Laws dispose of it, whenever the Welfare of the Publick shall demand such Disposition."[64] This statement may have said more than Franklin meant, but it reflected his outrage. Congress must have the power to tax for the public good, he believed.

Like Washington, Franklin held that independence required union and, without an effective central government, freedom was at risk. On May 13, 1784, one day after the formal exchange of ratified copies of the peace treaty following their return from the British and American capitals, Franklin sent a warning to Congress. "Our future Safety will depend on our Union and our Virtue. Britain will be long watching for Advantages, to recover what she has lost," he wrote. "If we do not convince the World that we are a Nation to be depended on for Fidelity in Treaties; if we appear negligent in paying our Debts, and ungrateful to those who have served and befriended us; our Reputation, and all the Strength it is capable of procuring, will be lost, and fresh Attacks upon us will be encouraged and promoted by better Prospects of Success."[65]

For Franklin, a people's virtue rested on their dedication not only to individual liberty and equality of opportunity, but also to loyalty, hard work, frugality, and paying their debts. For the United States, this required union. "If it had not been for the Justice of our Cause, and the consequent Interposition of Providence," he now wrote in a quasireligious explanation of the Revolution's success, "we must have been ruined."[66] This was how the Enlightenment era deist in Franklin thought: what is best for the individual naturally aligns with what is good for society. "Only a virtuous people are capable of freedom," he later observed. "As nations become corrupt and vicious, they have more need of masters."[67]

THE EXCHANGE OF TREATIES gave Franklin one more chance to press for retirement. John Jay and Henry Laurens, with whom he got along well, would soon return to America, and he asked them to plead his case. "I have reason to complain that I am so long without an Answer from Congress to my Request of Recall," Franklin wrote to Laurens in March 1784. "I wish rather to die in my own Country than here; and tho' the upper Part of the Building appears yet tolerably firm, yet being undermin'd by the Stone and Gout united, its Fall cannot be far distant. You are so good as to offer me your Friendly Services. You cannot do me one more acceptable at present, than that of forwarding my Dismission."[68] He made a similar plea to Jay in May: "Repose is now my only Ambition," Franklin stated.[69] The letter to Jay likely carried more weight than the one to Laurens because Congress had already tapped the New Yorker as its next secretary of foreign affairs. Still, it took time.

Along with their duties as peace negotiators, Jay and Laurens had been working with Franklin and Adams to open trade with various European nations. Upon the departure of Jay and Laurens, and seeking sectional balance in crafting trade deals, Congress created a three-member commission to negotiate treaties of friendship and commerce with the nations of Europe, north Africa, and Asia Minor. Jefferson, who worked well with both Franklin and Adams, replaced Jay and Laurens, giving each region of the country—north, middle, and south—one representative. This compromise meant once again denying Franklin's request to retire. Congress granted it only a year later, upon completion of the most pressing treaties. "You are permitted to return to America as soon as convenient," Jay wrote in a letter received by Franklin in May 1785. "This Circumstance must afford great Pleasure to your Family and Friends here."[70] It did.

As if to prove the comment in his 1784 letter about his mind being firm, during the period between the preliminary peace deal with Britain in November 1782 and his departure from France in July 1785, in addition to negotiating treaties of friendship with various nations and performing his duties as ambassador, Franklin resumed his pursuits in science and technology. In a remarkably original deduction, he attributed the unusually cold winter of 1783–1784 in Europe to the atmospheric impact of emissions from an Icelandic volcano. He also proposed the concept of daylight saving time, invented bifocals, played a role in the origins of human flight by hot-air and hydrogen-filled balloons, and served on a French royal committee investigating the medical claims of mesmerism.

When it came time for Franklin to leave Paris, the queen supplied a cushioned litter to bear him with minimal discomfort to the coast. Crowds gathered along the route for one last look at the man many considered the greatest savant of the age. His friends wept openly and ladies of the court begged him to stay. "The United States will never have a more zealous and more useful servant than Mr. Franklin," Vergennes noted.[71] Jefferson, Franklin's successor as American ambassador to France, heartily agreed.[72] On hearing that Franklin would return, Francis Hopkinson, a fellow signer of the Declaration of Independence then living in Philadelphia, wrote to him, "Whilst there is any Virtue left in America the Names Franklin & Washington will be held in the highest Esteem."[73] Upon learning of Franklin's arrival, Washington also wrote, "Amid the public gratulations on your safe return to America, after a long absence, and the many eminent services you have rendered it—for which as a benefited person I feel the obligation— . . . none can salute you with more sincerity, or with greater pleasure than I do."[74]

BY THE TIME FRANKLIN REACHED PHILADELPHIA in September 1785, Washington had been back at Mount Vernon for nearly two years. From the outset, Washington referred to it as his "seat of retirement from the bustle of the busy world."[75] Yet he had plenty to do there. Covering more than seven thousand acres, Mount Vernon was a vast business with hundreds of resident workers engaged in all sorts of plantation-related enterprises from planting through processing. "An almost entire suspension of every thing which related to my own Estate, for near nine years, has accumulated an abundance of work for me," Washington observed only days after his return.[76] He did not trust his slaves and regularly complained that they shirked work, stole supplies, and broke tools.[77] He felt a need to watch them daily to keep them on task. Sometimes, he would measure their output in his presence and then demand similar productivity during the entire workday, which lasted from sunrise to sunset with two hours for lunch, or up to fifteen hours per day, six days a week, in summer.[78] Washington often distrusted his hired overseers and paid workers as well and closely monitored their efforts. He was a hands-on manager by nature, but conditions at Mount Vernon accented this trait. "I made no money from my Estate during the nine years from it," he explained, now wanting to right the situation.[79]

Despite his formal retirement from public service and full-time occupation managing his plantation, Washington still carried the weight of a country on his shoulders. He had fought too hard to secure its independence not to care deeply about its survival, which he saw at risk under the Articles of Confederation. His earliest postretirement letters from Mount Vernon railed about Congress's lack of authority, failure to pay public creditors, and inattention to business. During the first half of 1784, he sent scores of letters to governors, former military colleagues, and members of Congress urging greater union. In a letter to the wartime governor of Connecticut, who had also just stepped down, Washington no sooner mentioned "the serenity of retirement" than he began grousing about the "deranged state of public Affairs."[80] To his own governor in Virginia he predicted nothing but "the worst consequences from a half starved, limping [central] Government."[81] Barely three months after returning to Mount Vernon, Washington conceded to Jefferson, "How far upon more mature consideration I may depart from the resolution I had formed of living perfectly at ease—exempt from all kinds of [public] responsibility, is more than I can, at present, absolutely determine."[82]

Washington's two postwar concerns—establishing his own estate and the United States—combined in his vision for the American west. Intent on securing his fortune in land, prior to the Revolutionary War, Washington had obtained large undeveloped tracts on the frontier in western Virginia and Pennsylvania. With peace, he sought to capitalize on his investment. Further, like Franklin, he viewed the west as key to America's future by being a source of both individual opportunity and economic expansion. Thus, after spending the first nine months of his so-called retirement trying to restore order to his plantation, Washington headed west to inspect his frontier holdings. This trip crystallized his hopes and fears for the country and began his journey back from retirement to full-time public service.

Washington traversed land that he knew from his days as a frontier surveyor and officer. Indeed, the outbound journey roughly followed Braddock's route toward the Forks of the Ohio, from which point Washington planned to journey down the Ohio Valley to his farthest and largest holdings. He never got that far. At his first large tract beyond the Allegheny Mountains, Washington found surly settlers who had leased the land from his agent but had little ability to pay their rent. At a second, larger tract, he found illegal squatters unwilling to recognize his ownership. He never reached seven farther tracts, including his largest ones on the Great Kanawha River in western Virginia, because Native peoples had reoccupied much of this area and, he was warned, warriors were lying in wait to capture or kill him. Virginia had ceded its claims to land northwest of the Ohio to the confederation, but Congress had no resources to defend or develop it. Worse still, the British had not evacuated their forts on the Great Lakes and still traded with and supplied guns to Native hunters and trappers. Settlers in the Ohio Country could turn toward the British in Canada or Spanish in Louisiana for protection and trade. Firsthand, Washington now confronted the consequences of a failing confederation and returned home ever more determined to address the problem.

As he saw it, the danger was not limited to territory northwest of the Ohio River but encompassed the entire frontier. "The Western settlers, (I speak now from my own observation) stand as it were upon a pivot," Washington wrote upon his return from the west, "the touch of a feather, would turn them any way."[83] Spain controlled the mouth of the Mississippi; Britain the St. Lawrence. He detected little loyalty to the United States or any individual state in the people whom he encountered on the frontier. "The ties of consanguinity which are weakening every day will soon be no bond," he warned.[84] "If then the trade of that Country should flow through the Mississipi or St Lawrence," Washington cautioned a member of Congress, "[i]f the Inhabitants thereof should form commercial connexions, which lead, we know, to intercourse of other kinds—they would in a few years be as unconnected with us, indeed more so, than we are with South America; and wd soon be alienated from us."[85] For the good of the country and his personal financial well-being, Washington concluded, America must secure the frontier. It offered another urgent argument for enhanced national power based squarely on military might, economic expansion, and imperial pretensions.

In the meantime, pending a fortified union, Washington initiated an ambitious effort to link Virginia to the frontier by improving navigation on the Potomac River as an alternative waterway to the west. The project would require dredging and canalling on a scale never before attempted in the states, but he offered to manage it personally as president of a private canal company. Within weeks of his return to Mount Vernon, Washington sent a shower of letters boasting of the profits that would flow from western navigation, warning of losing the west without it, and reporting on his observations that suggest its feasibility. Linking profit and patriotism, Washington hailed Potomac River navigation as "the cement of interest, to bind all parts of the Union together by indissoluble bonds—especially that part of it, which lies immediately west of us."[86] This was how the pragmatic providentialist in Washington thought: public policy should align individual self-interest with what is good for society as a whole.

For Washington, the presidency of the canal company became a consuming occupation, although one that he pursued while also managing his plantation and investment properties. Not only did he raise money through selling stock but he threw himself into deciding between cutting sluices through rapids or digging bypass canals around them, hiring supervisors and workers, and personally overseeing the glacial progress of reshaping a river with primitive tools. "Retirement from the public walks of life has not been so productive of the leisure & ease as might have been expected," Washington wryly remarked in a 1785 letter to the newly returned and supposedly retired Benjamin Franklin.[87]

The prospect of commercial traffic on the upper Potomac brought to the fore long-simmering jurisdictional disputes between Virginia and Maryland. By virtue of their colonial charters, Maryland claimed the river but Virginia controlled its southern bank. Under the Articles of Confederation, states were republics unto themselves. Each could have its own taxes and tariffs, rules and regulations, and currency, regardless of their impact on interstate commerce. Unless states cooperated, travel along an interstate boundary like the Potomac River could impose insuperable problems for people and products. To facilitate such cooperation, Virginia and Maryland appointed commissioners to address legal barriers to Potomac River commerce. Because their work impacted his company, Washington invited them to deliberate at Mount Vernon. A gracious and interested host who liberally lubricated his guests with good wine, he made sure that the commissioners reached agreement on critical matters of tolls, tariffs, and trade. They also agreed on shared funding for navigational aids, common fishing rights, and cooperation on protecting travelers. Breaking down barriers to interstate commerce, the commissioners reasoned, would benefit both states.

Known as the Mount Vernon Compact, it was quickly ratified by the legislatures of both states. James Madison served as floor manager for the compact in the Virginia assembly and received credit for its passage, even though Washington had a larger hand in crafting it. "We are either a United people, or we are not," he wrote to Madison shortly before the legislative debate began, and "if the former, let us, in all matters of general concern act as a nation."[88]

Building on this success and likely at Washington's urging, Madison persuaded the Virginia assembly to call for all the states to send delegates in September 1786 to a convention in Annapolis "to consider how far an uniform system in their commercial intercourse and regulations might be necessary to their common interest and permanent harmony."[89] At the time, trade disputes like those dividing Maryland and Virginia afflicted many states. Delaware, Pennsylvania, and New Jersey battled over their respective rights to use the Delaware River, for example, while New York, New Jersey, and Connecticut clashed over tariffs imposed by New York on goods passing through New York harbor on their way to or from other states.

Delegates from only five states showed up on time for the Annapolis Convention but they included Madison and Hamilton. Even before it began, both men recognized that any convention limited to commercial issues could not resolve the array of problems facing America, especially since, under the Articles of Confederation, Congress and all the states would need to approve whatever it did. Nothing but a complete overhaul of the Articles, drafted by an unbounded convention and ratified in a manner that would neither have to pass Congress nor require the approval of every state, could achieve the desired results. They wanted this first meeting to serve as a prelude to a second that (even before going to Annapolis) Madison had depicted as "a plenipotentiary convention for mending the Confederation."[90]

When the Annapolis meeting failed to attract delegates from enough states to proceed, Hamilton proposed that those present simply call for a second convention and go home. All told, the delegates met for three days. Their closing report, drafted by Hamilton and approved by the delegates on September 14, urged their states to "use their endeavors to procure the concurrence of the other States, in the appointment of Commissioners, to meet at Philadelphia on the second Monday in May next, to take into consideration the situation of the United States, to devise such further

provisions as shall appear to them necessary to render the constitution of the Federal Government adequate to the exigencies
of the Union."[91] This proposed conclave, its proponents hoped,
would be a true constitutional convention leading to the establishment of a true national government.

Some already charged that the Annapolis meeting could have
attracted more delegates and achieved more results if Washington
had participated, as he had for the Mount Vernon accords. The
challenge for federalists became getting him to Philadelphia in
1787 for the proposed grand convention. No one then knew if he
would—not even Washington. In the meantime, Pennsylvania,
Maryland, and Delaware pushed ahead with efforts to reduce trade
barriers and improve navigation in the lower Delaware and upper
Chesapeake Bays. In instructions to his state's delegates, Franklin embraced the spirit of compromise that would guide his own
actions at the Constitutional Convention. "The States," he wrote,
"have the same general objects, but as each may be attached to
ways of accomplishing them particularly favorable to itself, unless
a spirit of mutual concession take place among the Negociators,
a partial biass may tend to disappoint the main purpose." Therefore, he directed Pennsylvania's delegates to "sometimes yield in
points not materially disadvantageous to the State when it may be
necessary to procure a general concurrence."[92]

WASHINGTON WAS ONLY FIFTY-ONE and in good health (except for his
rotten teeth and recurrent headaches) when he resigned his military commission, so people reasonably expected him to do more
for his country, but Franklin was seventy-nine and physically infirm upon his retirement as ambassador. Repose seemed fitting.
Indeed, Washington wrote at the time about Franklin's "setting
himself down in the lap of ease, which might have been expected

from a person of his advanced age."[93] Accordingly, when the retired general wrote from Mount Vernon to welcome home the
aged patriot, he observed, "It would give me infinite pleasure to
see you: at this place I dare not look for it; tho' to entertain you
under my own roof would be doubly gratifying."[94] Washington
assumed any meeting between the two preeminent heroes of
the Revolution would have to occur in Philadelphia. Yet Franklin's mind was still sharp and he retained a youthful exuberance
and optimism. Washington in fact acted older than Franklin and
brooded more about death.[95]

Franklin did not waste a moment of his final transatlantic
crossing. He devised and used instruments to measure sea tem-
perature at various depths in order to determine that the Gulf
Stream, which he had "discovered" years earlier, operated as a
warm surface current over colder deep water. He tinkered with
his by then famous stove to reduce smoke when burning soft coal.
He worked on various ideas of watertight bulkheads, twin hulls,
and other innovations for ship safety, stability, and efficiency.
These and other shipboard researches by Franklin drew on his
seemingly instinctive understanding of energy, fluid, and air flow
and applied it to practical concerns. Commenting on the rigors of
an ocean voyage for a person of Franklin's age, Washington could
not but marvel, "A Man in the vigor of life could not have borne
the fatigues of a passage across the Atlantic, with more fortitude,
and greater ease than Doctor Franklin did."[96]

After arriving in Philadelphia following what he called "a pleasant Passage of 5 Weeks & 5 day," Franklin published his scientific
findings from the ocean voyage in the *Transactions* of the American Philosophical Society, an organization he founded in 1743
and now set about to revive and expand as the signal scientific
association of the new republic.[97] It had continued to elect him as
its president throughout his absence in France. Although focused

on the natural sciences, the society ventured into the mechanical arts, social sciences, and political theory.

Politics took a practical turn as well. No sooner had Franklin disembarked than each faction in Pennsylvania's bitterly divided assembly asked him to run as its candidate for the state's executive council from Philadelphia. One faction mainly represented the farmers and artisans, the other the commercial class. Both claimed Franklin, who won by acclamation. Under the state's constitution, which Franklin had largely written in 1776, members of the unicameral legislature and executive council annually elected the state's president by joint ballot from among the council's twelve members. Franklin then won that election too, with only one dissenting vote.

"He has again embarked on a troubled Ocean; I am persuaded with the best designs, but I wish his purposes may be answered—which, undoubtedly are to reconcile the jarring interests of the State," Washington noted at the time.[98] "If he should succeed, fresh laurels will crown his brow; but it is to be feared that the task is too great for human wisdom to accomplish."[99]

Washington underestimated Franklin. "He has destroyed party rage in our state," famed Philadelphia physician and former congressman Benjamin Rush observed late in Franklin's first term. "His presence and advice, like oil on troubled waters, have composed the contending waves of faction which for so many years agitated the State of Pennsylvania."[100] In 1786, Franklin won reelection unanimously. After a third election in 1787, term limits barred him from serving again. By the end of his tenure as his state's chief executive, Franklin was eighty-two.

Franklin's three terms as Pennsylvania's president spanned the nadir of the confederation period, but his state fared relatively well. The main problems of the era were economic, which fell to the states to address. Wartime and postwar spending on foreign goods had carried off much of America's hard currency, with the inability of Congress to impose protective tariffs or pay its wartime debts to domestic creditors worsening the situation. A deflationary spiral led to recessions in states like Massachusetts that did not stimulate spending by increasing the money supply, but doing so too aggressively by excessive issuances of unbacked paper money fueled chaotic inflationary cycles in states like Rhode Island and Georgia.

Before Franklin's return, Pennsylvania forged a clever middle course by unilaterally assuming the obligation to make interest payments on Congress's debts to in-state creditors through the issuance of negotiable IOUs (or indents), which creditors could use to pay state taxes. Although (unlike Rhode Island's widely disparaged paper money) not legal tender, these indents had sufficient intrinsic value to circulate like cash in Pennsylvania until used to pay taxes, thus enlarging the state's money supply in a sustainable way.[101] Once redeemed, Pennsylvania credited the indents toward payment of its annual requisition to Congress, which assured that its contributions went toward paying off Pennsylvania creditors rather than foreign debts or new expenses. Of course, Congress received no hard cash. Some other states soon implemented similar schemes.

In 1785, Congress tried to bar states from paying requisitions with indents, but Pennsylvania got around it by exchanging national securities held by its citizens for state bonds paying interest in the form of state tax credits. Under Franklin's leadership, it then fulfilled its requisitions to Congress by offsetting the interest paid or principal owed on these securities. The negotiable interest coupons on these state bonds continued to feed the in-state money supply in a stable fashion that kept the Pennsylvania economy growing without excessive inflation at a time when many other states suffered recessions or worse. All the while, Franklin boasted about his state's being one of few paying its full requisition. By 1786, Pennsylvania was Congress's single largest creditor.

Despite Pennsylvania's relative prosperity, Franklin knew that the ongoing political situation held back the country. Congress lacked power to address issues of general concern. "The Disposition to furnish Congress with ample Powers augments daily, as People become more enlightened," Franklin advised Jefferson early in 1786, adding a year later that the Articles of Confederation are "generally thought defective."[102] Regarding the defects in the Articles, or "errors" as he called them, Franklin wrote in November, "Those we shall mend."[103] After all, going back to his 1754 Albany Plan and subsequent efforts at the Second Continental Congress, Franklin had lobbied for an effective federal union longer than any living American.

Accordingly, under Franklin's leadership, Pennsylvania was one of the five states represented at the Annapolis Convention on interstate commerce. He approved when his state's delegate, Trench Coxe, backed the aborted conclave's closing call for a plenipotentiary constitutional convention in Philadelphia for May 1787. As Pennsylvania's president, Franklin would host it in his state's capitol building. Of course, this assumed that the Pennsylvania legislature would endorse the idea and appoint delegates to attend. That remained uncertain.

On December 1, 1786, after his state became the first to back the call for a plenary convention, Virginia governor Edmund Randolph dispatched two urgent letters to Franklin imploring Pennsylvania to act too. The first asked for Franklin's "co-operation in this trying moment." The second informed Franklin that Virginia had chosen its delegates to the Philadelphia Convention, and pleaded, "I have only to wish the presence of Pennsylvania by her deputies at this intended meeting." The call, Randolph assured Franklin, "breaths a spirit truly foederal and contains an effort to support our General Government, which is now reduced to *the most awful crisis.*"

Five days later, Randolph sent Franklin a third letter, which repeated his earlier request and stated that Washington would head the Virginia delegation. "My anxiety for the well being of the foederal government will not suffer me to risque so important a consideration upon the safety of a single letter," Randolph wrote to explain the multiple entreaties.[104]

Randolph need not have worried so much about Franklin's support for the Philadelphia Convention. Upon receiving his first letter, Franklin assured Randolph on December 21, "I communicated it to the Council, and it was sent down recommended to the Assembly. They took it into Consideration, and yesterday pass'd a Bill appointing seven Commissioners to meet yours in May next."[105] For health reasons, Franklin's name was not yet on the list, but soon joined it. Indeed, by April 1787, Franklin was hailing the pending convention as "an assembly of Notables" and cautioning Jefferson, "If it does not do Good it must do Harm, as it will show that we have not Wisdom enough among us to govern ourselves."[106]

For his part, Franklin was fully invested in the cause.[107] Despite his gout and gravel (as he termed his ailments), Franklin rarely missed any of the four-month-long marathon proceedings. Much as the crises of the French and Indian War and American Revolution drew Franklin and Washington together in the past, the crisis over union and reforming the confederation propelled them on converging paths toward the Constitutional Convention of 1787. As the delegates gathered in Philadelphia and word of the meeting spread among republican-minded European intellectuals, the English polymath Erasmus Darwin wrote to Franklin, "Whilst I am writing to a Philosopher and Friend, I can scarcely forget that I am also writing to the greatest Statesman of the present, or perhaps

Ambassador Benjamin Franklin, painted by Joseph-Siffred Duplessis in France, c. 1785.

President George Washington, one of some seventy-five "Athenaeum-type" portraits by Gilbert Stuart from a 1796 unfinished original.

Conjectural painting of Washington crossing the Allegheny River with Christopher Gist, who accompanied Washington on his 1753 mission to protest French occupation of the Ohio Country, by Daniel Huntington, mid-1900s.

Title page of the initial published version of Washington's report on his mission to the Ohio Country, 1754.

THE
JOURNAL
OF
MAJOR *George Wafhington,*

SENT BY THE
Hon. ROBERT DINWIDDIE, Efq;
His Majefty's Lieutenant-Governor, and
Commander in Chief of *Virginia,*

TO THE
COMMANDANT of the *French* Forces
ON
O H I O.

To which are added, the
GOVERNOR's LETTER:

AND A
TRANSLATION of the *French* Officer's Anfwer.

WITH
A New MAP of the Country as far as the
MISSISSIPPI.

WILLIAMSBURGH Printed,
LONDON, Reprinted for T. Jefferys, the Corner
of St. Martin's Lane.
MDCCLIV.
[Price One Shilling.]

Conjectural engraving of Washington conferring with his soldiers and Native American allies during the 1754 French assault on Fort Necessity at the outset of the French and Indian War by artist John McNevin, 1855.

Sketch illustrating Franklin's military efforts on the Pennsylvania frontier in 1756 during the French and Indian War from an 1849 edition of his *Autobiography*.

Earliest authenticated portrait of Washington, showing him as a colonel in the Virginia Regiment during the French and Indian War, painted by Charles Willson Peale, 1772.

Earliest known American political cartoon, designed by Franklin to urge colonial unity during the French and Indian War, from Franklin's *Pennsylvania Gazette,* May 9, 1754.

mitted to do, *muſt end in the Deſtruction of the Britiſh Intereſt,* Trade *and Plantations in America.*

JOIN, or DIE.

We hear that the General Aſſembly of this Province have voted the Sum of Ten Thouſand Pounds to be given to the King's Uſe at

Benjamin Franklin Drawing Electricity from the Sky by Benjamin West, c. 1816, commemorating the 1752 kite experiment.

(bottom left) Cover page for Franklin's *Poor Richard's Almanack* for 1748.

(bottom right) Title page for the constitution of the Pennsylvania Abolition Society, which was enlarged under Franklin's leadership in 1787.

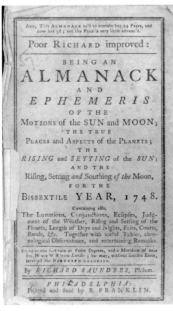

Poor RICHARD improved:

BEING AN

ALMANACK

AND

EPHEMERIS

OF THE

MOTIONS of the SUN and MOON;

THE TRUE

PLACES and ASPECTS of the PLANETS;

THE

RISING and SETTING of the SUN;

AND THE

Rising, Setting and Southing *of the* Moon,

FOR THE

BISSEXTILE YEAR, 1748.

Containing also,

The Lunations, Conjunctions, Eclipses, Judgment of the Weather, Rising and Setting of the Planets, Length of Days and Nights, Fairs, Courts, Roads, &c. Together with useful Tables, chronological Observations, and entertaining Remarks.

By RICHARD SAUNDERS, Philom.

PHILADELPHIA:

Printed and Sold by B. FRANKLIN.

[3]

THE

CONSTITUTION

OF THE

PENNSYLVANIA SOCIETY,

FOR PROMOTING THE

ABOLITION OF SLAVERY,

AND THE RELIEF OF

FREE NEGROES,

UNLAWFULLY HELD IN

BONDAGE;

ENLARGED AT PHILADELPHIA, APRIL 23d, 1787.

IT having pleased the Creator of the world, to make of one flesh all the children of men—it becomes them to consult and promote each other's happiness, as members of the same family, however diversified they may be, by colour, situation, religion, or different states of society. It is more especially the duty of those persons, who profess to maintain for themselves the rights of human nature, and who acknowledge the obligations of Christianity, to use such means as are in their power, to extend the blessings of freedom to every part of the human race; and in a more particular manner, to such of their fellow-creatures, as are entitled to freedom by the laws and constitutions of any of the United States,

and

Published text of the "Declaration of Independence" in Franklin's *Pennsylvania Gazette,* July 16, 1776.

(bottom left) Handwritten letter from Franklin to Washington, June 21, 1776. The two engaged in a lively correspondence throughout the Revolutionary era.

(bottom right) Handwritten letter from Washington to Franklin, December 28, 1778.

Declaration of Independence by John Trumbull, 1818, commemorating the presentation of the Declaration to Congress by the five-member drafting committee, including Franklin (*standing at right of center*).

Franklin's Reception at the Court of France, 1778, depicting Franklin receiving a laurel wreath upon his head, print by Anton Hohenstein, 1860s.

BENJAMIN FRANKLIN.

Né à Boston, dans la nouvelle Angleterre le 17 Janvier 1706

Franklin in France wearing his American bearskin cap, etching by Augustin de Saint-Aubin after Charles-Nicolas Cochin II, 1777.

Washington with his costumed slave William Lee on a bluff above the Hudson River during the American Revolution, painted from memory by the general's former aide-de-camp John Trumbull, 1780.

Conjectural painting of Washington and the Marquis de Lafayette at Valley Forge during the winter of 1777–1778 by John Ward Dunsmore, c. 1907.

(top left) Deborah Read
Franklin, attributed to
Benjamin Wilson, 1758.

(top right) Sarah Franklin Bache,
portrait by John Hoppner, 1793.

Martha Dandridge Custis
a year before her marriage
to Washington, portrait by
John Wollaston, 1757.

Conjectural illustration of Franklin greeted by his family and cheered by citizens on his return to Philadelphia in 1785 by J. L. G. Ferris, 1932.

The Washington family at Mount Vernon, showing George and Martha Washington; Martha's grandchildren George Washington Parke Custis and Eleanor Parke Custis; and one of Washington's slaves, probably William Lee, by Edward Savage, 1798.

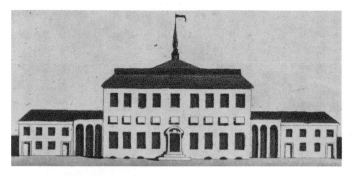

The State House in Philadelphia as it looked in 1787 during the Constitutional Convention.

U.S. Capitol mural showing delegates to the Constitutional Convention meeting with Franklin under the mulberry tree in his garden, painted by Allyn Cox, 1973.

Handwritten draft of a speech by Franklin to delegates at the Constitutional Convention.

Printed draft of the U.S. Constitution with Washington's handwritten notes, September 12, 1787.

WE, the People of the United States, in order to form a more perfect union, ■ eſtabliſh juſtice, inſure domeſtic tranquillity, provide for the common defence, promote the general welfare, and ſecure the bleſſings of liberty to ourſelves and our poſterity, do ordain and eſtabliſh this Conſtitution for the United States of America.

ARTICLE I.

Sect. 1. ALL legiſlative powers herein granted ſhall be veſted in a Congreſs of the United States, which ſhall conſiſt of a Senate and Houſe of Repreſentatives.

Sect. 2. The Houſe of Repreſentatives ſhall be compoſed of members choſen every ſecond year by the people of the ſeveral ſtates, and the electors in each ſtate ſhall have the qualifications requiſite for electors of the moſt numerous branch of the ſtate legiſlature.

No perſon ſhall be a repreſentative who ſhall not have attained to the age of twenty-five years, and been ſeven years a citizen of the United States, and who ſhall not, when elected, be an inhabitant of that ſtate in which he ſhall be choſen.

Repreſentatives and direct taxes ſhall be apportioned among the ſeveral ſtates which may be included within this Union, according to their reſpective numbers, which ſhall be determined by adding to the whole number of free perſons, including thoſe bound to ▬▬▬▬▬▬ for a term of years, and excluding Indians not taxed, three-fifths of all other perſons. The actual enumeration ſhall be made within three years after the firſt meeting of the Congreſs of the United States, and within every ſubſequent term of ten years, in ſuch manner as they ſhall by law direct. The number of repreſentatives ſhall not exceed one for every forty thouſand, but each ſtate ſhall have at leaſt one repreſentative: and until ſuch enumeration ſhall be made, the ſtate of New-Hampſhire ſhall be entitled to chuſe three, Maſſachuſetts eight, Rhode-Iſland and Providence Plantations one, Connecticut five, New-York ſix, New-Jerſey four, Pennſylvania eight, Delaware one, Maryland ſix, Virginia ten, North-Carolina five, South-Carolina five, and Georgia three.

When vacancies happen in the repreſentation from any ſtate, the Executive authority thereof ſhall iſſue writs of election to fill ſuch vacancies.

The Houſe of Repreſentatives ſhall chooſe their Speaker and other officers; and they ſhall have the ſole power of impeachment.

Sect. 3. The Senate of the United States ſhall be compoſed of two ſenators from each ſtate, choſen by the legiſlature thereof, for ſix years: and each ſenator ſhall have one vote.

Immediately after they ſhall be aſſembled in conſequence of the firſt election, they ſhall be divided as equally as may be into three claſſes. The ſeats of the ſenators of the firſt claſs ſhall be vacated at the expiration of the ſecond year, of the ſecond claſs at the expiration of the fourth year, and of the third claſs at the expiration of the ſixth year, ſo that one-third may be choſen every ſecond year: and if vacancies happen by reſignation, or otherwiſe, during the receſs of the Legiſlature of any ſtate, the Executive thereof may make temporary appointments until the next meeting of the Legiſlature, which ſhall then fill ſuch vacancies.

No perſon ſhall be a ſenator who ſhall not have attained to the age of thirty years, and been nine years a citizen of the United States, and who ſhall not, when elected, be an inhabitant of that ſtate for which he ſhall be choſen.

The Vice-Preſident of the United States ſhall be, ▬▬▬▬, Preſident of the ſenate, but ſhall have no vote, unleſs they be equally divided.

The Senate ſhall chooſe their other officers, and alſo a Preſident pro tempore, in the abſence of the Vice-Preſident, or when he ſhall exerciſe the office of Preſident of the United States.

Washington during the Constitutional Convention, painted and engraved by Charles Willson Peale, 1787.

Franklin during the Constitutional Convention, painted and engraved by Charles Willson Peale, 1787.

Howard Chandler Christy's monumental 1940 *Scene at the Signing of the Constitution of the United States,* showing Washington presiding and Franklin at center.

Detail of *The Apotheosis of Washington* from the apex of the U.S. Capitol's rotunda (*Washington at top center and Franklin at lower left*), fresco by Constantino Brumidi, 1865.

any century[,] Who spread the happy contagion of Liberty among his countrymen."[108] In America and Europe, many friends of liberty (as they liked to call themselves) looked on the Convention as the next stage for establishing a working model for republican rule. The presence there of Franklin and Washington gave reason to hope for transformative results—and ultimately proved essential to realizing them.

Book III

WORKING TOGETHER AND APART

Six

RENDEZVOUS IN PHILADELPHIA

WITH CIVIC BOOSTERS INTENT on turning Philadelphia into the commercial, financial, and perhaps political capital of the newly independent states, the city looked better than ever by the spring of 1787. Residents had repaired damage caused by the British occupation and begun fresh construction. No longer hampered by prewar British restrictions on foreign trade and wartime harassment of American shipping, trade routes opened to Europe and the Caribbean. In 1785, the *Empress of China,* a three-mast sailing ship built during the war as a privateer and owned by a syndicate led by Philadelphia merchant Robert Morris, returned from China with the first direct commerce between the world's newest republic and its oldest empire. Washington bought a set of porcelain tableware carried in the ship.

With peace, Morris had resigned as superintendent of finance for the Confederation but remained a power in state politics and America's richest merchant prince. Washington depicted Morris's three-and-a-half-story Georgian mansion, located only a block from the State House and boasting a half-acre walled garden, as the finest single residence in Philadelphia. Home to the resident governor prior to the Revolution and the British commander during the occupation, Morris bought and expanded it in 1781. Pennsylvania's pro-federalist assembly picked Morris, who had led the failed fight to gain taxing authority for Congress, as a delegate to the Constitutional Convention.

Two blocks east of Morris's mansion on Market Street, Franklin had replaced three run-down, street-front rental buildings with two four-story row houses (one containing his ground-floor printing shop) and an archway through to his home. Also since returning from France, he had added to his house a three-story wing featuring a first-floor dining area seating twenty-four and a massive second-floor library and scientific study.

All this might seem improvident for someone of his age, Franklin acknowledged in a 1786 letter to his last surviving sibling, "but we are apt to forget that we are grown old, and building is an amusement."[1] That "amusement," he noted on the eve of the Constitutional Convention in 1787, involved supervising "Bricklayers, Carpenters, Stonecutters, Plaisterers, Painters, Glaziers, Limeburners, Timber Merchants, Coppersmiths, Carters, Labourers, etc. etc."[2] In assuming this task, Franklin had in mind his daughter, son-in-law, and seven grandchildren, who all resided with him. Although a doting grandfather, Franklin confessed to enjoying the quiet of his new library where, as he put it, "I can write without being disturb'd by the Noise of the Children."[3] Writing to his sister with the house complete, he marveled at his good fortune, "When I look at these Buildings, my dear Sister, and compare them with that in which our good Parents educated us, the Difference strikes me with Wonder."[4]

The fifty-year-old State House had changed too. The rotting steeple that towered over the symmetrical, Georgian-style structure when Washington, Franklin, and others had met there for the Second Continental Congress came down in 1781, replaced by a squat temporary cupola that would remain in place for five decades. New landscaping, paving, fencing, and copper spouts improved the exterior. The landscaping, which included a public garden with a grove of elms, serpentine walks, and constructed landforms, so delighted Washington that he asked its architect to create similar features at Mount Vernon. A city hall, courthouse, and home for Franklin's American Philosophical Society were under construction on the State House grounds. With the confederation Congress then meeting in New York, the Convention could convene in the historic chamber where the Continental Congress had declared independence, without risking interference from the current government. No city in America had more rooming houses and taverns to lodge and feed delegates. Theaters and private clubs offered evening entertainment.

Philadelphians prepared for the coming fifty-five-member conclave. In February, long before the first out-of-state delegate arrived, they formed an elite Society for Political Inquiries that met biweekly in the library of Franklin's home to discuss matters of political economy and public responsibility in a republic. Every member of the Convention from Pennsylvania joined, as did many of the civic leaders who would host delegates at salons and dinners. Although the society avoided topics of governmental structure that might divide members and intrude on the Convention's business, it took on ones that framed and gave meaning to republican government. In May, the society made Washington an honorary member.[5]

As state president, Franklin also arranged for the Convention to meet in the spacious, ground-floor Assembly Room of the State House rather than nearby Carpenters' Hall, which had housed the First Continental Congress, and he otherwise helped to prepare Philadelphia for the coming conclave. Reading about his activities in the newspapers of far-off Boston, Franklin's sister wrote to him with evident pride that it "makes you Apear to me Like a young man of Twenty-five." Her prayer for the Convention was simple: "I had Rather hear of the Sword being beat into Plow-Shares," she told her brother, "if by that means we may be brought to live Peaceably with won a nother."[6]

OF ALL THE DELEGATES descending on Philadelphia, none was more anticipated than Washington. His inspirational leadership during the Revolution—serving without pay or leave for more than eight years—followed by his acclaimed retirement from public life at the war's end made him the personification of republican virtue at a time when Americans hungered for a leader they could trust with power. On April 3, Franklin wrote to Washington expressing his "Hopes of seeing you here at the Convention, being persuaded that your Presence will be of the greatest Importance to the Success of the Measure."[7] This was the only personalized greeting extended by Pennsylvania's sitting chief executive. Robert Morris soon added the invitation "that you will on your arrival come to our House & make it your Home during your Stay in this City. We will give you as little Trouble as possible."[8] The Morrises offered this courtesy only to Washington. Other delegates lodged in crowded rooming houses and dined at common tables in them or at taverns and clubs. In May, Pennsylvania's vice president wrote to Franklin, "I believe it will be Proper for [the Executive] Council to Address Genl Washington. Your Excellency knows best what should be done."[9] No other delegate received a formal state welcome.

As Franklin's letter suggests, Washington's attendance at the Convention remained in doubt until the last minute. The prior December, Virginia had tapped him to lead its delegation to Philadelphia, but he did not initially accept and instead urged that someone "on whom greater reliance can be had, may be substituted in my place."[10] Virginia governor Edmund Randolph entreated Washington "not to decide on a refusal immediately."[11]

James Madison wrote separately to express his "wish that at least a door could be kept open for your acceptance hereafter, in case the gathering clouds should become so dark and menacing as to supercede every consideration, but that of our national existence."[12] Those considerations included Washington's concerns that Congress had not yet consented to calling the Convention; the fact that states might so restrict their delegations that they could not (as Washington put it) "probe the defects of the Constitution to the bottom, and provide radical cures"; and disabling rheumatoid arthritis in his shoulder, which might keep him from traveling.[13] In short, beyond matters of health, Washington did not know whether the convention route could work and did not want to expend his limited political capital on a doomed enterprise.

One by one, Washington's concerns evaporated or at least became manageable. In February, Congress consented to calling the Convention "for the sole and express purpose of revising the Articles of Confederation."[14] This measure might not go as far as Washington wished toward authorizing the Convention to craft radical cures for what he saw as a fatally flawed confederation but in April he heard from Secretary of War Henry Knox. Every state except Rhode Island would send delegations to the Convention, Knox wrote, and none would carry unduly limiting instructions on what its delegates could discuss or propose.[15]

Acting separately, Knox, Madison, and Foreign Secretary John Jay also sent Washington outlines for a new constitution. All called for turning the current governing structure on its head by replacing a toothless unicameral congress composed of delegates from sovereign states with a sovereign national government comprised of a bicameral legislature, independent executive, and separate judiciary. The authority of Congress should expand to cover "All national objects" as Congress itself defined them, Knox noted.[16] Jay wrote of "the States retaining only so much power as may be necessary for domestic Purposes."[17] Madison called for a federal system "which may at once support a due supremacy of the national authority, and not exclude the local authorities whenever they can be subordinately useful."[18]

Impressed by the similarities in these three proposals, Washington prepared an abstract comparing their main features that he could use as a strategic plan for the Convention. Much as on the eve of battles during the American Revolution, when he would listen to his top officers before framing a plan of operations, Washington finally felt ready to proceed. He now had a plan, or at least an outline of a plan, for a truly national government.

So long as his health allowed, Washington told Randolph at the end of March, he would go to Philadelphia.[19] There he would deliberate with the assembled state delegations, he explained to Jay, to find out "what can be effected." The Philadelphia Convention, Washington here added in an expression of urgency surpassing any of his prior warnings, "may be the last peaceable mode" of saving the union.[20] Already first in war by military service, Washington would attempt by political service to become first in peace. Encouraging Washington to assume this role, Knox assured him that, should the Convention produce a wholly new constitution rather than "patch work to the present defective confederation," he would be "doubly entitle[d] . . . to the glorious republican epithet—The Father of Your Country."[21] It was at this point that Washington received the letter from Franklin expressing the Pennsylvanian's view that Washington's presence at the Convention would be vital to its success.[22]

For Washington, as the reasons against attending the Convention faded, motives for going appeared in sharper relief. Among the dark clouds impelling him forward, thousands of rural Massachusetts debtors had taken up arms in a desperate effort to stop state courts from foreclosing on their farms during the economic depression engulfing their state in 1786. Unlike Franklin's Pennsylvania, Massachusetts had done little to counter the destabilizing postwar constriction in the money supply, which the confederation left each state to address. Loosely led by a former Continental Army captain and wounded veteran of the Battle of Saratoga named Daniel Shays, this local insurrection appeared more frightening the further removed observers lived—and Washington resided four states away. He learned about it mostly through Knox.

"The fine theoretical government of Massachusetts has given way," Knox wrote to Washington after rioters stopped courts from sitting in six rural counties during the September 1786 term. Despairing of the states, Knox concluded, the general federal "government must be braced, changed, or altered to secure our lives and property."[23] After the closures extended into the December term, Washington wrote back, "I feel, my dear Genl Knox, infinitely more than I can express to you, for the disorders which have arisen in these states. Good God! who besides a tory [loyalist] could have foreseen, or a Briton predicted them!"[24]

In 1787 as in 1776, Washington equated liberty with the defense of property rights, which helped to make slavery such a hard issue for him and led him so ready to denounce Shays and his fellow "Regulators," despite their sympathetic status as unpaid war veterans. "Notwithstanding the boasted virtue of America, we are far gone in every thing ignoble & bad," he complained to Knox.[25] Accordingly, Washington hailed the response of Massachusetts governor James Bowdoin, whose harsh debt and tax policies had fueled the uprising, to hire a twenty-five-hundred-man private army to pursue these Regulators and punish their leaders. "If government shrinks," Washington wrote as Bowdoin's army set out, "fresh manœuvres will be displayed by the insurgents—anarchy

& confusion must prevail—and every thing will be turned topsy
turvey."[26] Bowdoin did not trust the regular Massachusetts militia
to disperse the Regulators, especially since some of its soldiers had
joined forces with the insurgents.

In the defense of property rights, Washington found an ally in
Franklin. After Bowdoin's army routed the ill-equipped Regula-
tors in February 1787 and sent their leaders fleeing out of state,
Franklin, as president of Pennsylvania, threw the full weight of
his state's government behind efforts to apprehend the runaways.
"I congratulate your Excellency most cordially on the happy
Success attending the wise and vigorous Measures taken for the
Suppression of that dangerous Insurrection," he wrote to Bowdoin
after signing legislation adding to the reward offered for Shays's
capture.[27] The voters of Massachusetts were less impressed. They
booted Bowdoin out at the next election and reinstated as governor the more moderate John Hancock, who pardoned Shays and
promoted debt and tax reform. The crisis had passed in Massachusetts but it left a lingering impact on the movement for constitutional reform. "These disorders are evident marks of a defective
government," Washington wrote to the Marquis de Lafayette in
March about the episode; "indeed the thinking part of the people
of this Country are now so well satisfied of this fact that most of
the Legislatures have appointed, & the rest it is said will appoint,
delegates to meet at Philadelphia the second monday in may next
in general Convention of the States to revise, and correct the defects of the federal System."[28]

FINALLY, IN EARLY MAY, the last impediment to Washington's attendance at the Convention fell away as his health stabilized enough
to travel. To others, Washington had always appeared a tower of
physical and emotional strength, especially when on horseback:

tall, solidly built, reserved, immaculately dressed, and a superb equestrian. The clothes and the emotional reserve remained but, at fifty-five, he no longer displayed quite the same physical force. Acquaintances commented that he looked older and somewhat stooped. Rheumatism in his shoulder could make riding painful and caused him at times to carry his arm in a sling. After delaying his departure due to the rain that dogged his entire trip, Washington left for Philadelphia by carriage on May 9 still complaining of a "violent" headache and "sick stomach."[29] His only traveling companions were three slaves—his valet Billy Lee, coachman Paris, and groom Giles—perched on front or back of his carriage. Having become "too attentive to two little Grand Children to leave home," Washington explained, his wife remained behind at Mount Vernon.[30] These grandchildren were hers, not his. She had taken responsibility for rearing them after their father died. They were heirs to the vast Custis estate, much of which Martha held in trust.

A reminder of early America's expansiveness, Washington's 150-mile carriage journey from Mount Vernon to Philadelphia took five days. It also crossed two culturally, economically, and religiously distinct regions by going from the hierarchical, slave-holding south with its plantation-based economy and entrenched Anglicanism to the more egalitarian middle states of small family farms, growing commercial centers, and deep religious diversity. Fully representative of the former region, Washington, the beneficiary of inherited wealth and status by birth and marriage, saw himself as part of the gentry while from the latter, Franklin, who earned his wealth and status, always portrayed himself a "middling" or middle-class man.[31] By the time of the Constitutional Convention, Franklin had freed his few house slaves and Pennsylvania had decreed the gradual abolition of slavery within its borders, while Washington retained roughly three hundred enslaved

workers at Mount Vernon and the number of slaves throughout
Virginia continued to grow. The differences dividing Franklin's
Pennsylvania and Washington's Virginia, which only increased as
one traveled to states farther north or south, illustrated the task
faced by delegates seeking to forge a true federal union out of
thirteen far-flung states from three distinct regions: northeast (or
east), middle, and south.

Not having announced his departure date or travel route in
advance, Washington's arrival in each community along the way
came as something of a surprise until he neared Philadelphia.
This suppressed what would have otherwise been a triumphant
procession. Word had reached Philadelphia in advance, however,
and an all-star delegation waited to greet him at the state line in
Chester on May 13. Nearer Philadelphia, mounted dragoons of
the city light horse brigade took their place around Washington's
carriage as it passed in paradelike fashion between rows of uniformed troops and crowds of local citizens. "On my arrival, the
Bells were chimed," Washington noted in his diary.[32] The Convention was scheduled to begin the next day—May 14, 1787.

Washington first went to the boardinghouse operated by Mary
House, where Madison had taken up residence, and then a few
doors west to his own lodgings at Morris's mansion. Minutes
later, he emerged to travel some three hundred yards to Franklin's home where, as Washington wrote, he "Waited upon the
President."[33] This was Washington's first order of business—his
first formal act—upon reaching Philadelphia. The leading delegate calling on the host state's chief executive: the nation's two
most celebrated heroes conferring and, by doing so, giving dignity to the proceedings that brought them together. Although
Washington likely walked, historian of the Convention Richard
Beeman envisions him traveling by carriage because the "meeting
with Franklin had the character of a formal state visit." Various

accounts have Franklin showing Washington his house and the two men getting reacquainted over tea or wine under the shade of a mulberry tree in the courtyard. "That garden meeting may well have provided America's two most illustrious statesmen with the opportunity to form a bond that would prove immensely valuable in the months to come," Beeman posits.[34] One of the sixteen historical murals in the U.S. Capitol's Great Experiment Hall depicts Franklin conferring at this site with leading delegates during the Convention.

THE CONVENTION GOT OFF TO A SLOW START. On Monday, May 14, Washington and presumably Franklin appeared at the State House at the appointed hour for the Convention to start but found only Madison and other delegates from Pennsylvania present. They returned the following two days, but no more states were represented, even though four more members from Virginia arrived. The Convention could not meet without a quorum of at least seven states represented by half or more of their delegates.

Late on the afternoon of the sixteenth, Washington returned to Franklin's house for what amounted to a state dinner for those delegates already in attendance. "We have here at present what the French call *une assemblée des notables,* a convention composed of some of the principal people from the several states of our confederation," Franklin wrote to the London brewer who had supplied the beer used on this occasion. "They did me the honor of dining with me last Wednesday, when the cask was broached, and its contents met with the most cordial reception and universal approbation. In short the company agreed unanimously that it was the best porter they had ever tasted." Franklin hosted a similar dinner during the next week. Despite his advanced age and well-known infirmities, he was clearly up for the Convention and a prime mover at the proceedings. "When I consider how many terrible diseases the human body is liable to, I comfort myself that only three incurable ones have fallen to my share, viz, the gout, the stone, and old age; and that these have not yet deprived me of my natural cheerfulness, my delight in books and enjoyment of social conversation," Franklin added in his letter to the brewer.[35]

Given the primitive state of medicine at the time and the average forty-four-year life span for a white American in 1790, Franklin was in tolerably good health for a man his age. Although once a fine horseman and expert swimmer, riding and swimming were now out of the question and even standing or walking caused pain from his kidney stone. "Sitting, or lying in Bed I am generally quite easy," Franklin wrote to a friend in France less than a month before the scheduled start of the Convention, "and as I live temperately, drink no Wine, and use daily the Exercise of Dumb Bell, I flatter myself that the Stone is kept from augmenting." To minimize the pain, when Franklin acted in his official capacity as Pennsylvania's president (which likely was stretched to include attending the Convention), inmates from the nearby Walnut Street Prison sometimes bore him the three blocks from his home to the State House on a sedan chair. "People who live long, who will drink of the Cup of Life to the very Bottom, must expect to meet with some of the usual Dregs," he wrote to this French friend.[36] Franklin was fifteen years older than the Convention's second most senior member, Roger Sherman, and twice the average age of the others yet, as one delegate noted, possessed "an activity of mind equal to a youth of 25."[37]

Most important, Franklin remained eternally optimistic and enthusiastic. Only weeks before the Convention began, with Shays's Rebellion and the problems of paper money in Rhode Island likely in mind, he wrote to other friends in France, "Our public affairs go on as well as can reasonably be expected after

so great an overturning. We have had some disorders in different parts of the country, but we arrange them as they arise, and
are daily mending and improving; so that I have no doubt but all
will come right in time."[38] To Jefferson in Paris, he reported about
the pending Convention, "The Delegates generally appointed as
far as I have heard of them are Men of Character for Prudence
and Ability, so that I hope Good from their Meeting."[39] Franklin
wished Jefferson could be there too but carried on without him,
advancing many of the positions that the younger Virginian likely
would have championed.

In contrast, Franklin probably was relieved that John Adams
stayed put in Europe during the Convention. The two had fallen
out during their time in France and Adams had not written to
Franklin for nearly a year when, early in 1787, Adams sent him
unsolicited copies of his hastily written treatise on American constitutions, which the vainglorious New Englander hoped might
guide the deliberations in Philadelphia. The treatise favored too
much executive and senatorial power for Franklin's taste. He sent
back a bland acknowledgment only days before the Convention
began in a chilly letter that reserved its warmest words for Adams's wife and daughter.[40]

A lifelike portrait from the period by Charles Willson Peale
shows a clear-eyed but clearly aged Franklin with a deep double
chin, balding crown, bifocals, and stringy, shoulder-length hair.
Heavy rain kept Franklin home on the Convention's first day, but
he never missed another session, giving him the best attendance
record of any Pennsylvanian and, among all delegates, second
place only to Washington and Madison, who attended every day.
Beeman counts these three as the Convention's *only* indispensable
members.

Making the most of the delay caused by lack of a quorum,
Washington, Franklin, and their fellow delegates from Virginia

and Pennsylvania formulated a plan for going forward. Beeman credits Franklin's inaugural dinner party for launching the process. "Franklin," he wrote, "had a superb sense of the way in which good food, liquor, and conversation could lubricate the machinery of government and politics, and his dinner gathering was designed for that purpose."[41] Historian Catherine Drinker Bowen has the Virginians meeting alone in the mornings and with the Pennsylvanians each afternoon.[42] Beeman does not differentiate.

The meetings between delegates from these two key states began soon after Franklin's May 16 dinner—convening for "two or three Hours, every day," according to Virginia delegate George Mason. This group expanded Madison's outline for the Constitution into what became known as the Virginia Plan.[43] In a letter on May 21, Mason depicted the emerging plan as "a total Change of the Federal System, and instituting a great national Council, or Parliament upon the principles of equal proportionate Representation, consisting of two Branches of the Legislature, invested with full legislative Powers upon the Objects of the Union; and to . . . establish also a national Executive; and a judiciary System."[44] In short, the draft created a true national government with a two-house legislature and a balance of powers among its branches. Assuming that the entire Pennsylvania and Virginia delegations attended some or all of these sessions, then Franklin and Washington were present at the creation of the modern American government. They certainly supported the essential plan when it came before the full convention, although Franklin favored a single-house legislature like the one in Pennsylvania, a system that later became common in northern Europe.[45]

ON FRIDAY, MAY 25, nearly two weeks after its scheduled start, the Convention finally gained a quorum. With just seven southern and middle states represented, the delegates could do little more than open the proceedings and prepare for future deliberations. Principally, that meant electing a presiding officer. As president of the host state, Franklin planned to nominate Washington, but this was the day that rain kept the Pennsylvanian from attending what was sure to be a largely ceremonial session. The task of nominating Washington fell to the state's next most senior delegate, Robert Morris, who pointedly stated that he did so on behalf of his entire delegation. South Carolina's John Rutledge seconded the nomination and urged that it be accepted unanimously. It was.

Morris and Rutledge then escorted Washington to the State Assembly Speaker's chair: a finely carved seat with a half sun painted on its crown that would serve as the presiding officer's perch for the convention. It stood behind a draped desk on the dais at the front of the Assembly Room.

After sitting, Washington accepted the "honor," as he put it, of presiding, noted the "novelty" of the president's role for him, and begged "the indulgence of the House toward the involuntary errors which his inexperience might occasion." He had not asked to preside but surely anticipated doing so. If "he felt himself embarrassed" with the honor, as he stated at the time, it was a fleeting feeling.[46] Comfortable with command, Washington was better suited to direct the proceedings from in front than engage in debate from the floor.

Three more brief business items came before the Convention on that initial day. First it chose William Jackson, a former aide of Washington, as Convention secretary. This disappointed Franklin, who hoped that the post would go to his grandson Temple, who clearly would have done a better job than Jackson, who took poor notes and threw away most of the records. Then one member from each state presented the credentials for his delegation. Finally, before adjourning for the week, the Convention named a committee chaired by Washington ally George Wythe to draft standing rules for the assembly.

In presenting the credentials from his state, George Read made a point of stating that Delaware had barred its delegation from supporting any change in the policy of equal representation for every state in Congress. This caught the attention of every delegate. It appeared in all their surviving notes.[47] Mason promptly wrote home, "Delaware has tied up the hands of her deputies by an express direction to retain the principle . . . of each State having the same vote."[48] An avowed federalist, Read favored a strong general government but wanted his small state to have as much say in it as any large one. His opening remark suggested that the chief question at this convention would not be if Congress received more power but rather whether Congress would represent states or people. Both Franklin and Washington backed proportional representation but that stance raised the added question: Who counts as people? Obstacles loomed ahead.

FRANKLIN MADE HIS INITIAL APPEARANCE at the Convention on Monday, May 28, along with enough delegates from Massachusetts and Connecticut to bring the total number of states represented to nine. Although the comings and goings of delegates meant that not every state remained represented at all times, the total never again dropped below eight or rose above twelve. About thirty of the seventy-five elected delegates showed up on a typical day; twenty of them never appeared. Franklin arrived on his first day by sedan chair, which likely added to his aura. "Dr. Franklin is well known to be the greatest phylosopher of the present age," Georgia delegate William Pierce wrote from the Convention. "The very heavens obey him, and the Clouds yield up their Lightning to be imprisoned in his rod."[49]

The delegates began the new week by debating draft rules proposed by Wythe's committee. These rules placed substantial authority in Washington, as president, to manage the Convention. They also provided for the delegates to vote by state—one vote for each state as cast by a majority in its delegation—with a majority of states voting needed to pass any motion or measure.

The most significant rule did not come from the committee but was offered by South Carolina's Pierce Butler. He moved from the floor that the Convention proceed in secret. The final rules provided that "no copy be taken of any entry on the journal" and that "nothing spoken in the House be printed, or otherwise published, or communicated without leave."[50] These rules allowed the Convention to build internal consensus without outside interference. "No Constitution ever would have been adopted by the Convention if the debates had been public," Madison later commented.[51] Albeit incomplete, Madison's extensive handwritten notes of the debates, later revised and published, provide the best record of what happened behind closed doors.[52]

Washington scrupulously followed the secrecy rule in public discourse and private writings. "No Com[municatio]ns without doors," he wrote in his diary for May 29, never again disclosing details of the deliberations.[53] When one delegate accidentally breached the secrecy rule by mislaying a copy of the Virginia Plan, Washington's stern supervision of the deliberations showed itself. "I am sorry to find that some one Member of this Body, has been so neglectful of the secrets of the Convention as to drop in the State House a copy of their proceedings," Washington lectured the delegates after another member found it. "I must entreat Gentlemen to be more careful, lest our transactions get into the News Papers, and disturb the public repose by premature speculations." He then threw down the offending document, directed its unnamed owner to claim it, and stormed from the chamber. "It is something remarkable that no Person ever owned the Paper," Georgia's Pierce commented.[54]

WITH SHUTTERS CLOSED, observers excluded, and secrecy enforced, the main business of the Convention began on Tuesday, May 29, with Washington calling on Governor Randolph to introduce and defend the Virginia Plan, which took the form of fifteen numbered resolutions. "In a long and elaborate speech," New York antifederalist Robert Yates reported, Randolph "candidly confessed" that the intent behind these resolutions was to create "a strong *consolidated* union, in which the idea of states should be nearly annihilated."[55]

Yates exaggerated the plan's intent, but not by much. Under the Articles of Confederation, each state was sovereign; the general government was not. "The State of Georgia, by the Grace of God, free, sovereign and independent," the credentials for that state's convention delegates stated.[56] The Constitution altered this equation. Before the Constitution, for example, Georgia printed its own paper money. After ratification, it could not. Before, it could limit imports from other states. After, it could not. Before, it could maintain its own militia. After, the president could nationalize those troops. Before, the federal government could not meddle with Georgia's peculiar institution of slavery. After . . . well, that was where federalists from Georgia and the Carolinas drew a line, and those from Virginia largely supported them. At the Convention, Franklin and Washington generally stood shoulder to shoulder on consolidating federal authority, but issues like slavery divided them and their federalist allies into camps, leading to some key compromises over federalism.

In the framing of the Constitution, the Virginia Plan constituted the federalists' opening salvo, but it required elaboration and was subject to concessions. To forge a general government responsible for the "common defense, security of liberty and general welfare," it gave Congress unfettered power "to legislate in all cases to which the separate States are incompetent, or in which the harmony of the United States may be interrupted by the exercise of [state] Legislation."[57] This Congress would consist of two branches, with the first elected by the people of the several states proportional to their free population or tax payments and the second chosen by the first. Congress would choose the chief executive for a single fixed term and judges for life. Congress could strike state laws for violating the Constitution and a council composed of judges and the executive could void congressional acts on the same grounds.

Franklin and Washington might quibble with bits of the Virginia Plan, but it set forth the sort of government they sought. "The business of this Convention is as yet too much in embryo to form any opinion of the result," Washington wrote to Jefferson at this time. "That something is necessary, all will agree; for the situation of the General Governmt (if it can be called a governmt) is shaken to its foundation—and liable to be overset by every blast. In a word, it is at an end, and unless a remedy is soon applied, anarchy & confusion will inevitably ensue."[58] Yet even if *all* agreed that *something* was necessary, not everyone agreed on that something. "The players of our game are so many," Franklin would later say about the Convention, "their ideas so different, their prejudices so strong and so various, and their particular interests independent of the general seeming so opposite, that not a move can be made that is not contested."[59] A passionate chess player, Franklin concluded that at times the moves and countermoves of the delegates in Philadelphia "confound the understanding."[60]

IN ADDITION TO countless lesser addendums and alterations, turning the Virginia Plan into the framed Constitution required three major compromises or innovations, each of which engaged Franklin and Washington. Foremost among these, the so-called Great Compromise restructured Congress to have a proportionally representative, popularly elected House of Representatives (like the lower branch of many state assemblies) and a Senate with equal representation from each state (akin to the confederation Congress but with its state-appointed members serving fixed terms). Although Franklin personally favored a popularly elected unicameral legislature, he foresaw the final compromise from the outset and did as much as anyone to broker it.

While a majority of delegations supported proportional representation for Congress, a determined minority, mostly from small states, demanded equal representation and threatened to scuttle any deal without it. Although in theory a mere quorum of the delegations could proceed, and a mere majority of that quorum could frame a constitution, every delegate knew that any working general government would need to include all of the large states—slave and free—and most of the small ones. The United States might get along without Rhode Island, which did not attend the Convention, but it would not be the United States without Pennsylvania, Virginia, Massachusetts, New York, South Carolina, and most of the rest. They held leverage at the Convention and, motivated by some mix of high ideals and self-interest, they used it.

Seeking middle ground on the issue of representation in Congress, delegates from the midsized state of Connecticut offered the obvious compromise of proportional representation in Congress's lower house and equal representation in the Senate, but such a mishmash only slowly won out.[61] A fight over principle with practical implications, the contest over representation was not finally resolved until late July.

After simmering for weeks, the floor debate erupted into a verbal brawl on Saturday, June 30. "Can we forget for whom we are forming a Government? Is it for *men,* or for the imaginary beings called *States?*" Pennsylvania's James Wilson asked from a large-state viewpoint.[62] He objected to a majority of the states having power to set the nation's course when those states contained a minority of its people and wealth: they could easily abuse their power through how they imposed taxes. Others saw their rights and welfare flowing from the states and did not want to disenfranchise them. Arguing from the small-state perspective, New Jersey's Jonathan Dayton dismissed Connecticut's proposed compromise as "an amphibious monster" that his people would never accept.[63]

Taking the measure of both sides in his folksy, pragmatic way, Franklin now broke with his state to argue in favor of the compromise. "If a proportional representation takes place, the small States contend that their liberties will be in danger. If an equality of votes is to be put in its place, the large States say their money will be in danger," he stated. Noting that "we are all met to do something," Franklin urged the Convention to act like a carpenter who, when framing a table from two planks of uneven parts, "takes a little from both, and makes a good joint."[64]

At this point, some large-state delegates wanted to call the bluff of those demanding equal representation. "If a minority should refuse their assent to the new plan of a general government," Wilson asserted, it could not happen on better grounds.[65] Madison agreed. Gunning Bedford of Delaware shot back, "The Large States dare not dissolve the confederation. If they do the small ones will find some foreign ally . . . who will take them by the hand and do them justice." Turning to the large-state delegates, he said with emphasis, *"I do not, gentleman, trust you."*[66] The Convention adjourned for the day in disarray with a vote on the compromise scheduled for Monday, July 2, 1787.

WATCHING THIS DEBATE FROM THE CHAIR, Washington nearly lost hope. Only three weeks earlier, he had written privately to his nephew George, "The sentiments of the different members seem to accord more than I expected they would," but now he doubted it.[67] On Sunday, Washington conferred with Pennsylvania's Gouverneur and Robert Morris. According to an early account of that meeting, all three were dejected by the "deplorable state of things at the Convention." They complained of conflicting opinions "obstinately adhered to" and members threatening to leave. "At this alarming crisis," the account noted, "a dissolution of the Convention was hourly to be apprehended."[68] Firmly in the large-state camp, Washington favored proportional representation in Congress but, like Franklin, was now willing to take from both sides to save the middle.

On July 2, when delegates finally voted on the so-called Connecticut Compromise, the Convention deadlocked: five to five with one state split. Virginia and Pennsylvania voted no. Declaring "we are now a full stop," Connecticut's Roger Sherman backed a motion to commit the matter to a committee, where cooler heads might prevail. Led by Madison and Wilson, many large-state delegates spoke against the move but, perhaps influenced by his meeting with Washington, Gouverneur Morris endorsed it, as did Washington's Virginia ally Edmund Randolph. The motion passed with Virginia and Pennsylvania voting yes.[69] The delegates then stacked the committee with moderates like Franklin from large states and hard-liners like Bedford from small ones. The committee's outcome was predetermined by its composition.

Franklin worked his mediating magic in committee. Hosting its members for dinner that same night and then meeting with them at the State House on the next day, Franklin proposed his version of the Connecticut Compromise. For small states, he offered "an equal vote" for every state in the Senate; for large states, he offered one member in the House of Representatives "for every forty thousand inhabitants" plus the concession that all bills raising or spending money would originate in this lower house and could not be amended by the Senate.[70] Presumably, this arrangement would protect the small states' liberty and the large states' money. Although each side objected to some part of the compromise, the committee agreed to submit it to the Convention. After more bitter debate and some minor amendments, the Convention narrowly passed the compromise on July 16 and never looked back. Franklin's greased language left unresolved how the limits on money bills would operate and who constituted "inhabitants" for purposes of apportioning House seats. In resolving these matters, Washington played a conciliating role.

The concession on money bills unexpectedly split the large-state delegates. Although it was intended to limit the power of the small-state-dominated Senate, some big-state delegates like Wilson, Madison, and Gouverneur Morris feared that the provision would undermine that body's ability to check democratic excess in the lower house. In August, after the bar against the Senate originating or amending money bills was included in a draft constitution compiled from provisions already approved by the Convention, these delegates pushed to remove it. Other large-state delegates viewed the concession as essential to their states' interests. "To strike out the section, was to unhinge the compromise of which it made a part," Virginia's George Mason complained.[71] Franklin agreed, adding that he welcomed having money bills crafted by the people's house.[72]

Casting the deciding vote in a divided Virginia delegation,

Washington first voted to strike the provision but, when the issue came up again five days later, switched sides and carried Virginia with him. "He disapproved & till now voted against," Madison wrote, but "gave up his judgment, he said, because it was not very material weight with him & was made an essential point with others, who if disappointed, might be less cordial in other points of real weight."[73] Ultimately, the Convention settled on compromise language that pleased Washington: "All Bills for raising revenue shall originate in the House of Representatives: but the Senate may propose or concur with amendments as on other bills."[74]

THE COMMITTEE'S PROPOSAL to apportion the House of Representatives based on the number of a state's "inhabitants" inadvertently opened the explosive issue of slavery, leading to a second major compromise. In early June, when the delegates originally debated the principle of proportional representation for Congress as set forth in the Virginia Plan, they were thinking of proportionality in terms of either the number of a state's free people or the amount of its tax payments, with tax payments serving as a rough gauge of property. The Convention had papered over the difference by simply agreeing to allocate seats "according to *some* equitable ratio of representation."[75] Franklin believed this ratio should factor in both people and property since government was instituted to protect both.[76] This left an opening for him to make some concession for slave property even though he opposed slavery. Representing a slave state, Washington also wanted concessions for slave property.

As a practical matter, because they had more slaves and fewer free people than their northern counterparts, southern states would lose representation under any allocation based only on free inhabitants. Likewise, they would gain under one that included all inhabitants, free or slave. The problem with the latter approach was, as some northern delegates noted, that the south treated slaves as property, not people. Although this was common practice throughout America in colonial times, the revolution for liberty had transformed popular thinking on the issue, at least in the north. If people were the sole basis for representation, Massachusetts's Elbridge Gerry asked when this matter first arose at the Convention, "Why then should the blacks, who were property in the South, be in the rule of representation?"[77] Gerry opposed slavery and wanted southerners to confront the hypocrisy of their position.

In mid-June, when delegates were still thinking in terms of an "equitable ratio" that factored in property as well as people, they accepted a plan proposed by Pennsylvania's Wilson and seconded by South Carolina's Charles Pinckney to apportion congressional seats among the states in proportion to the whole number of their free inhabitants plus "three fifths of all other persons," meaning slaves.[78] In adapting this now shocking formula from an earlier one proposed by Congress for equitably allocating the requisitions due from each state, the delegates were not thinking of slaves as having three-fifths of the moral worth of free persons but, in even less human terms, as having three-fifths of the property value of free people. Astonishing as it seems today, the so-called Three-Fifths Compromise passed with only Gerry speaking against it.[79] Even Gouverneur Morris, who opposed counting slaves as people because it would reward slaveholders, accepted doing so as a rough gauge of property.[80] Conceding the slave states anything less than this three-fifths, most delegates believed, would derail the Convention and, by various rationalizations, most northerners accepted it.

By mid-July, after the delegates agreed to Franklin's language of allocating House seats based on population with one representative for each forty thousand individuals, no one could disguise the three-fifths rule as merely a means to factor in the relative worth of southern property. "Individuals" could only mean people. With tensions heightened by the battle over Senate representation, other northerners now joined Gerry in arguing that slaves should not count because their states treated them as property. "If Negroes are not represented in the states to which they belong, why should they be represented in the General Congress?" New Jersey's William Paterson asked in a rhetorical question addressed directly to Washington.[81] South Carolina's delegation countered by demanding that slaves count as whole people, not just as three-fifths of one, despite their lack of any rights. "The security that the Southn States want is that their negroes not be taken from them," Pierce Butler thundered.[82] Virtually no one expressed concern for the slaves. They became white men's pawns in a north-south power struggle. Franklin remained quiet during this phase of the debate reflecting his belief that, even if it meant accepting the Three-Fifths Compromise, the best way to end slavery in southern states was to first get them in a strong federal union.

For those delegates principally concerned with preserving what had been gained to this point and saving the Convention from shipwreck, the goal became getting the delegations back to the Three-Fifths Compromise. Washington played his part when, in the midst of this debate, a committee chaired by Gouverneur Morris proposed an initial allocation of House seats. Attacked by delegates who thought their states underrepresented, Morris explained that his committee used an estimate—really "little more than a guess," he said—of population and property in allocating seats.[83] Washington promptly created another committee, which made new allocations using the Three-Fifths Compromise, and the crisis passed.[84] Three-fifths for their slaves was enough to keep the southern delegations from bolting without losing the northern ones. The Carolina and Georgia delegations would later win two additional constitutional protections for slavery, with Washington's Virginia supporting one, aiding the capture of fugitive slaves, and opposing the other, barring Congress from limiting the importation of slaves until 1808.[85]

Despite counting as three-fifths of a person for purposes of representation, slaves could not vote in any state—every delegate agreed on this, even those who hated slavery. This did not settle the question of which free citizens could vote in House elections. Franklin wanted as broad a franchise as possible, although he never said if this should include women and Native Americans. His Pennsylvania constitution guaranteed voting rights to all freemen. In neighboring New Jersey, the 1776 state constitution extended voting rights to all adult property owners, including women. Because state law barred married women from owning property, this construction restricted female suffrage to single women. The Virginia constitution, in contrast, limited voting rights to male landowners and many of Washington's allies at the Convention, including Madison and Gouverneur Morris, favored extending that limit to federal elections in all states.

"We shd. not depress the virtue & public spirit of our common people," Franklin shot back against this proposal. "The sons of a substantial farmer, not being themselves freeholders, would not be pleased at being disfranchised." He gave as a principle that the elected should not "narrow the privileges of the electors."[86] For New Jersey, this would mean preserving the voting rights of single women. Responding to Madison's worry that the poor could not be trusted with the vote, Franklin replied, "Some of the greatest rogues he was ever acquainted with, were the richest rogues."[87] The Convention split the difference by confirming the compromise that, in each state, the qualifications for voting in House elections would be the same as for voting in state assembly elections. Pennsylvania, New Jersey, Virginia, and every other state could follow its own light, no matter how dim. Free Blacks voted in some states, but not others, depending on local practice. As for Native Americans, the Constitution expressly excluded "Indians not taxed" from counting toward representation in Congress and thus they were barred from voting everywhere.[88]

In a similarly pragmatic move made near the Convention's end, when some delegates urged that the limit on the number of House members per state be lowered from no more than one for every forty thousand inhabitants to no more than one for every thirty thousand, Washington sided with them. "It was much to be desired that the objections to the plan recommended might be made as few as possible," he stated, and this change accommodated a general insecurity "for the rights & interests of the people."[89] With Washington's endorsement, the amendment passed. Enlightenment pragmatists at heart, both Franklin and Washington sacrificed lesser aims to achieve their principal goals.

HAVING BRIDGED THE GULF over representation dividing small versus large and free versus slave states with pragmatic (but unprincipled) compromises, the Convention faced the enigma of the executive. Revolutionary era Americans saw the face of tyranny in the likeness of King George III and feared creating a despot like him. But they had few places to turn for precedent in crafting the presidency of an extended republic. Each state had an executive officer, but their authority varied. Some, like Franklin in Pennsylvania, had little formal power and gained influence only by the deference they commanded. Others, like George Clinton, Washington's wartime friend and postwar business partner who served seven terms as governor of New York, wielded real power under his state constitution. Governors, unlike kings, did not have authority over foreign affairs or war and peace, however, and their jurisdictions were relatively small. Looking to Europe for examples of elected leaders with such powers, delegates cited the consuls of ancient republican Rome, the Holy Roman Emperor, the Venetian Republic's doge, the king of Poland, and even the Pope, but none of the analogies fit. The American presidency was the Convention's most original creation, with Franklin and Washington splitting over the extent of its powers.

The debates on the executive consumed more time at the Convention than those on any other topic and were not resolved until September. Having agreed to begin by working through the Virginia Plan, the delegates reached its two resolutions regarding the executive on June 1. The longer of these called for a "National Executive" chosen by Congress for a single term of some fixed but unspecified length. "Besides a general authority to execute the National laws," it stated, this officer "ought to enjoy the Executive rights vested in Congress by the Confederation." The shorter one provided a limited means of vetoing bills passed by Congress.[90]

If these executive rights included all those once held by the British monarch and later vested in Congress, the provisions gave considerable power in the executive. Beyond executing laws, the king held direct authority over war and peace, the military, foreign affairs, appointing officers and judges, and granting pardons. Since the Articles of Confederation vested power over these matters in Congress, they might go to the executive under the Virginia Plan. Then again, they might not. The resolutions were frustratingly vague on this score.

Perhaps because Washington was sitting among them, when the delegates reached these resolutions, they fell unusually silent. After brief comments by two supporters of a strong executive, Madison wrote in his notes, "a considerable pause ensued" and the chair asked whether the delegates were ready to vote on (and presumably pass) the provisions.[91] Coming from Washington's delegation, no one seemed inclined to dispute them.

Most delegates assumed Washington would become the first president, trusted him in that post, and wanted to shape the office to his satisfaction.[92] From the war, he had a popularity that bordered on veneration. Indeed, during the war, patriot propagandists promoted him as a republican replacement for the king. "That *George* is now no more," one such writer proclaimed soon after the Declaration of Independence, "GOD save great WASHINGTON."[93] By 1778, references to his being the Father of His Country begin appearing much as kings stood as father figures for their realms. By the 1780s, Americans celebrated his birthday in a manner once reserved for the king's. In 1784, Princeton College commissioned Charles Willson Peale to paint a portrait of Washington to replace (in the same frame) one of King George II for its main hall. A generation later, early-national-period author Washington Irving (whose parents named him for the general in 1783) had his late-colonial idler Rip Van Winkle awake after the Revolution to find King George's image on a tavern sign transformed to one of Washington simply by changing the coat's color from red to blue, swapping a sword for a scepter, and adding a cocked hat. So readily could one George replace another in American eyes, Irving seemed to say, so naturally would the later one become president, if not king.

Perhaps more important, in his person, Washington projected an aura that inspired deference. It fulfilled his aspiration, expressed in 1776 after taking command of the American army: "to obtain the applause of deserving men, is a heart felt satisfaction—to merit them, my highest wish."[94] A popular account has Hamilton, at the Convention, warning Gouverneur Morris that Washington was too aristocratic and reserved for friends to treat as they did others. Dismissing Hamilton's remark,

Morris bet that he could greet Washington with a gentle slap on the shoulder. When Morris did so, the story goes, "Washington withdrew his hand, stepped suddenly back, fixed his eye on Morris for several minutes with an angry frown, until the latter retreated abashed." Morris claimed his wager but said of the episode, "Nothing could induce me to repeat it."[95] Showing the regard with which delegates held Washington, this account suggests why they might defer to his wishes on the presidency.

But who would follow Washington in that post?

Fully Washington's equal and never one to defer, Franklin broke the silence at the Convention over the presidency. Emphasizing that the structure of the executive is "of great importance," he urged delegates to "deliver their sentiments on it before the question was put."[96] This comment burst the dam and debate flooded the room. Four days later, with the discussion still raging, Franklin said with reference to Washington and the presidency, "The first man, put at the helm would be a good one. No body knows what sort may come afterwards. The executive will always be increasing here, as elsewhere, till it ends in a monarchy."[97]

Favoring a weak executive, at one point or another during the debates Franklin advocated circumscribing the presidency with term limits and an advisory council, and by eliminating the veto power. "In free Governments the rulers are the servants, and the people their superiors," he said of term limits. "For the former to return among the latter was not to *degrade* but to *promote* them."[98] Further, Franklin wanted presidents elected by the people and to serve without any compensation beyond their expenses. "I am apprehensive," he explained, "that the Government of these States, may in future times, end in Monarchy. But this Catastrophe I think may be long delayed, if in our proposed system we do not sow the seeds of contention, faction & tumult, by making our posts of honor, places of profit."[99] Twice, Franklin defended the power to impeach corrupt presidents against those like Gouverneur Morris who would put them above the law during their terms in office. "It wd. be the best way," Franklin explained, "to provide in the Constitution for the regular punishment of the Executive when his misconduct should deserve it, and for his honorable acquittal when he should be unjustly accused."[100]

By the second day of their discussion of the executive, the delegates had drifted so far apart that they could not even agree whether there should be one president or an executive triumvirate like those of late republican Rome, with a member from each of the country's three regions. With Washington in the room, a solo executive should have seemed obvious, especially since every delegate who knew him well must have known that he would never serve as one member of an executive committee. Fearful of investing too much power in any one person, some delegates— including Mason and Randolph from Washington's own state— favored a triumvirate. On June 1, Randolph denounced a solo executive as "the fetus of monarchy."[101] By the second, Franklin was echoing the Virginia governor's warning about "nourish[ing] the fetus of a King."[102] He later expressed the added concern that a solo executive might lead to disruptive discontinuity in government upon a change in administration or if the president became incapacitated. "The Steady Course of public Measures is most probably to be expected from a Number," he stated with respect to having an executive triumvirate or council.[103] In a vote taken that day, the delegations split six to four, with the majority not yet ready to accept a solo presidency.

As much as he enjoyed life at the Morris mansion, where he could dine every day, Washington often ate with other delegates at the common tables of taverns and public houses. Later that afternoon, after the extent of disagreement over the power, structure, and selection of the executive became apparent, he ate at City

Tavern, where the subject of that day's debate surely remained on everyone's mind. While in session that day, the delegates had raised and could not resolve the issue of whether the United States should have one executive officer or three. Now, as many of those members dined with the man who would be king, Washington's presence must have reassured them. One frequent guest at City Tavern, Pierce Butler, later commented that powers vested in the chief executive under the Constitution would not "have been so great had not many of the members cast their eyes toward General Washington as President."[104] At the Convention's next session, the delegations voted by a margin of seven to three for a single executive. Virginia joined the majority, with Washington casting the deciding vote within its five-member delegation.[105] Indeed, on every occasion during the proceedings, he cast his vote within the Virginia delegation for a stronger presidency.

AFTER WORKING SIX DAYS A WEEK for more than two months, on July 28, the Convention referred all the provisions passed thus far to a five-member Committee of Detail charged with organizing them into a single coherent document, and recessed for nine days. The delegates needed a break. Washington visited his old encampment at Valley Forge and went fishing with Gouverneur Morris. Having gradually lost stamina and become noticeably tired, Franklin recuperated at home. "It must be no small comfort for you to have a short resting spell," his grandson Benny wrote to him on August 1. "I really think your illness was in great measure owing to the fatigue you suffered while [the Convention] was sitting, but hope this respite from *that* business, will fortify your health."[106] Apparently it did. The surviving record suggests that Franklin's vigor flagged in late July before rebounding in August. At the Convention's end, he wrote, "Some tell me I look better, and they suppose the daily Exercise of going and returning from the State house, has done me good."[107]

As submitted to the full Convention on August 6, the committee's draft Constitution presented a snapshot of where matters stood at the time. The president, elected by Congress for a single seven-year term, would execute the nation's laws, possess a limited veto over legislation, hold the pardon power, and serve as commander in chief of the armed forces.[108] This executive would be beholden to Congress, which expressly held the formerly monarchical powers of declaring war and making peace.[109] The Senate, with two members appointed by each state's legislature for six-year terms, would act as a coequal branch of Congress in lawmaking plus hold the traditionally executive powers of making treaties and appointing ambassadors and judges and the judicial power of resolving disputes between states. In effect, it would have mixed legislative, executive, and judicial functions. The electorate of each state would choose members of the House of Representatives—a purely legislative body—in numbers proportionate to the state's free population and "three fifths of all other Persons."[110] Beyond creating a Supreme Court, the structure and powers of the judiciary were left intentionally vague, in large part because delegates raised but never settled whether courts could review the constitutionality of federal laws or if there would be lower federal courts.

These provisions provided the starting point for concluding deliberations on the separation of powers between Congress and the executive, which extended into September as weary delegates raced toward adjournment. During this final stage of the Convention, with Washington's apparent backing, the presidency gained power at the Senate's expense.[111] Franklin, who would have preferred giving more power to the popularly elected House of Representatives, largely stayed out of the tug-of-war between the Senate and executive, except to second a failed motion to have the president restrained by a regionally representative executive council such as he had as president of Pennsylvania. "A Council would not only be a check on a bad President," Franklin argued to no avail, "but a relief to a good one."[112] Far from curbing the executive, by this point the delegates (led by Washington allies Hamilton, Wilson, and Gouverneur Morris) seemed intent on augmenting its power.

The resulting changes in the power balance among branches, which laid the foundation for the American presidency, began with concerns over the Senate. Some delegates thought that the Committee of Detail, by allocating executive and judicial powers to the Senate, created an aristocracy, which Mason defined as "government of the few over the many."[113] Others soured on the Senate after it became the agency of the states rather than proportionally representative of the nation as a whole. Madison foresaw it perpetuating the failings of the confederation Congress by favoring state over national interests.[114] By late August, most delegates wanted to rein in the Senate, leaving open the question how to reallocate its broad powers. Those powers could have gone to Congress as a whole, or to the Supreme Court, but supporters of a strong executive saw their chance and, given the widespread trust in Washington as president, pounced.

While the members accepted most of the committee's draft Constitution, they deferred action on several key provisions relating to the presidency and Senate. Reaching an impasse, on August 31 they referred these postponed parts, which included basic matters regarding presidential selection and executive power, to a committee with one member from each state. Led by Madison and Gouverneur Morris, this committee revived an idea floated earlier by Wilson of having the president elected by separately chosen, state-based electors, who collectively became known as the Electoral College.[115] Freeing the presidency from selection by

Congress led the committee to increase its power.[116] Most important, the committee proposed that the president (rather than the Senate) make treaties and choose ambassadors and judges subject only to the Senate's advice and consent. With delegates anxious to finish, the Convention accepted these fundamental changes with little debate.[117]

So long as Congress elected the president, the delegates had limited the president to a single long term. Otherwise, they feared the executive would come under the sway of the legislators who could reelect him. Using independent electors to select the president opened the door for multiple terms.[118] Accordingly, the Convention settled on a four-year term for the president but no limits on reelection. With this final shift, the Convention struck its ultimate balance between the executive and Congress, giving birth to the American presidency with its vast authority and independent selection process. Most Western democracies would not follow this model, favoring instead prime ministers beholden to the legislature. Instead, powerful, independent presidents would become the hallmark of postmonarchical authoritarian regimes.

Surveying the final product evolved from their Virginia Plan, Edmund Randolph and George Mason warned fellow delegates that such a Constitution "would end either in monarchy, or a tyrannical aristocracy," and voted against it.[119] They found a vocal ally in Massachusetts's Elbridge Gerry. Franklin had expressed similar concerns throughout the proceedings but, in part due to his trust in Washington as the first president, endorsed the final draft. His lingering worries, however, may account for his widely quoted answer to the grand dame of Philadelphia high society, Elizabeth Powel, when she allegedly asked him after the Convention ended if it had created a republic or a monarchy. "A republic," Franklin reportedly replied, "if you can keep it."[120]

WITH RANDOLPH AND MASON NOW IN OPPOSITION, on September 15, when the delegates finally voted on the finished Constitution, Washington again cast the deciding ballot within his state's five-member delegation. His aye carried Virginia, and with it unanimous consent from the eleven states remaining at the Convention to send the document forward to Congress and the states for ratification. Without the support at this point of Washington's Virginia, prospects for ratification would have dimmed.

As it was, with Randolph, Mason, and Gerry voting no, and with New York having pulled its delegation, the resistance took form. In the fight over ratification, antifederalists would oppose the president's broad authority and the expansive list of federal powers that included not only to tax and spend for the general welfare and regulate interstate commerce but also to make any laws necessary and proper to implement the listed powers. In the give-and-take at the Convention, this list, which included every power sought by Washington, replaced the Virginia Plan's vague assertion of federal authority "in all cases to which the separate States are incompetent."[121] Antifederalists also decried the refusal of Washington, Madison, and other federalists to include a bill of rights. Too much presidential and federal power coupled with too few restraints put individual liberty and private property at risk, they argued, just as federalists maintained that too little of the former and too many of the latter courted those same dangers.

Riddled with compromises and jury-rigged provisions (somewhat like the war effort Washington once led), at core this was Washington's Constitution, especially with respect to the presidency. During the ratification campaign, he declared to its opponents, "In the aggregate, it is the best Constitution that can be obtained in this Epocha."[122] In particular, Washington defended the powers given to Congress as no more "than are indispensably necessary to perform the functions of a good Government"[123] and never doubted the broad authority conferred on the president even after Lafayette, writing from France early in 1788, singled out those "Extensive powers of the Executive" as one of only four points (along with no bill of rights, guarantee of jury trials, or presidential term limits) questioned by the European philosophers who had reviewed the document. Jefferson, he noted, concurred in these four concerns.[124] In his reply, Washington gave ground only on a bill of rights and guaranteed jury trials by suggesting that, in due course, amendments could provide for them.[125]

Franklin shared antifederalists' concerns over presidential power and wanted a more democratic Constitution but endorsed the final draft as better than nothing and perhaps the best of all. "I agree to this Constitution with all its faults," he told his fellow delegates in a major prepared address delivered on the Convention's final day, "because I think a general Government necessary for us, and there is no form of Government but what may be a blessing to the people if well administered" (or a bane, he implied, if administered poorly). Franklin understood the divisions splitting the Convention and opted to support whatever compromise could produce a workable federal government. If the states met again, he darkly warned, it would only be "for the Purpose of cutting one anothers Throats."[126]

With a nod to Washington, Franklin expressed his faith that the Constitution "is likely to be well administered for a course of years," yet predicted that it would "end in Despotism, as other forms have done before it, when the people shall become so corrupted as to need despotic Government." A Constitution providing more popular control and less executive power could better withstand corruption, Franklin believed, but "the older I grow, the more apt I am to doubt my own judgment, and to pay more respect to the judgment of others." Accordingly, he favored ratifying the Constitution. "The opinions I had of its errors, I sacrifice to the public good," he pledged. "Within these walls they were born, and here they shall die."[127] In short, while not his ideal Constitution, Franklin embraced it.

"It is a singular Thing in the History of Mankind," Franklin wrote to a friend in France, "that a great People have had the Opportunity of forming a Government for themselves."[128] Washington made similar comments to friends at home and abroad.[129] Both viewed the Convention as a modern triumph of reasoned debate and compromise in politics that could serve as a model for others. In October, for example, the ever forward-looking Franklin sent a copy of the Constitution to Rodolphe-Ferdinand Grand, who had managed French loans to the United States during the American Revolution. "I do not see why you might not in Europe carry on the Project of . . . forming a Federal Union and One Grand Republick of all the different States and Kingdoms by means of a like Convention," Franklin suggested, "for we [too] had many interests to reconcile."[130]

Franklin and Washington embraced the Constitution because it realized their long-held ambition for a fortified federal government with consolidated authority over commerce, defense, and taxation. Washington also secured a strong, independent, and unitary presidency that Franklin saw as overly monarchical. Coming from large states and fundamentally national-minded, neither Franklin nor Washington favored a Senate with two members from each state but both accepted it as a necessary compromise. Southern delegates (including Washington) also scored critical safeguards for slavery that many northern delegates (including Franklin) hoped would fail. It tells much about their rational pragmatism and faith in republican virtue that, despite its compromises, Franklin and Washington so unreservedly accepted the American Constitution.

ALTHOUGH WRITTEN BY A COMMITTEE and approved by the Convention, the cover letter transmitting the Constitution to the confederation Congress was signed solely by George Washington. "The friends of our country have long seen and desired, that the power of making war, peace, and treaties, that of levying money and regulating commerce, and the corresponding executive and judiciary authorities should be *fully* and effectually vested in the general government," the letter stated. Effecting these ends, it noted, justified "the *consolidation* of our Union, in which is involved our prosperity, felicity, safety, perhaps our *national* existence."[131] This letter launched the public campaign for ratification. Newspapers widely reprinted it along with Franklin's closing speech, which (alone of all the deliberations at the closed-door Convention) leaked to the press.[132] Taken together, these documents made it appear that the Constitution came directly from Washington and Franklin.

As the two larger-than-life leaders whose support made the Convention and Constitution credible, Washington and Franklin dominate the monumental historical painting of the event by Howard Chandler Christy that a later Congress commissioned for the United States Capitol. The shutters symbolically open and drapes pulled to reveal a bright new day that backlights the figures in an almost holy aura, Franklin sits at center surrounded by the other signers. Washington, standing alone, surveys the scene from an overly elevated dais.

Christy invented the arrangement of characters and tinkered with the setting, but his painting's spirit rings true. Washington did oversee the signing from his seat on the dais. Franklin had written the day's key address. The shutters remained closed throughout the proceedings, but the room may have seemed brighter than before on this final day. Although observed by no one except the Convention's members and officers, the signing likely felt as historic to them as it looks in Christy's painting. Washington signed first and above the rest in a bold, large hand somewhat reminiscent of John Hancock's already well-known signature on the Declaration of Independence: "G°: Washington, Presid^t and Deputy from Virginia." Then the other thirty-eight signers filed forward state by state, north to south, with Franklin in the middle leading the Pennsylvanians.

While the last members were signing, Franklin looked at the half sun adorning the crown of Washington's chair. "I have," he said to those near him, "often in the course of the Session, and the vicissitudes of my hopes and fears as to its issue, looked at that behind the President without being able to tell whether it was rising or setting: But now at length I have the happiness to know that it is a rising and not a setting Sun."[133] Madison chose this anecdote involving the Sage of Philadelphia, Washington's chair, and the rising sun of American nationhood to close his record of the Constitutional Convention. After relating it, Madison's notes simply conclude that, once the final member signed, "The Convention dissolved itself."[134] Washington left Philadelphia the next day pleased with the Convention's achievement. The draft Constitution, he soon made clear, fulfilled his hopes for the type of radical cures to the confederation's infirmities that had drawn him to Philadelphia in the first place.[135] Now he would fight for its ratification.

Two months later, the assembly and council of Pennsylvania unanimously reelected Franklin to another term as the state's president. "This universal and unbounded confidence of a whole People," he confessed to his sister Jane, "flatters my Vanity much more than a Peerage could do."[136] Given his devotion to public service, Franklin viewed his election as something akin to a republican ennoblement. It also gave him an official platform to promote ratification, which he hoped would happen quickly with his state leading the way.

Seven

DARKNESS AT DAWN

DESPITE THE MIRACLE AT PHILADELPHIA, as Washington later characterized the Convention's work, ratification remained in doubt.[1] Large and small, north and south, slave and free—the states would need to overlook vast differences to unite under a single Constitution. The process involved separate ratifying conventions in each state, with the approval of nine required for the document to take effect for those states. "The arguments . . . most insisted upon, in favor of the proposed constitution, are; that if the plan is not a good one, it is impossible that either General Washington or Dr. Franklin would have recommended it," one Pennsylvania commentator wrote in late October.[2] "Let the People confirm what was done by FRANKLIN the sage, and by brave WASHINGTON," a South Carolina essayist added a week later.[3] Even though neither Franklin nor Washington formally participated in the ratification process, as these newspaper articles suggested, each played a key part in it. Indeed, for their role in framing the Constitution, another correspondent had already hailed them jointly as "the fathers of their country."[4]

Both men, having taken center stage at the Convention, chose to operate behind the curtain in the ratifying process. As his state's president, Franklin declined to serve at Pennsylvania's ratifying convention even though partisans put his name forward. As the presumptive first president of the United States—a post he did not seek but everyone assumed he would take—Washington might have seemed self-serving at his state's convention. In addition, while able legislators, neither was a gifted public speaker and both favored the role of brokering compromises over facing down critics. Given their preeminence in Philadelphia, neither could hope to maintain a low profile at a heated state ratifying convention.

Their decision to keep above the fray made sense. Franklin and Washington were thin-skinned and recoiled from ad hominem attacks. Neither was particularly adept at defending a proposal in a contentious up-or-down vote, especially a proposition they helped to craft. Franklin might begin fiddling with the text and Washington might lose his temper. Each could be baited by opponents, and the Constitution would surely face fierce opposition at the ratifying conventions in Philadelphia and Richmond. Federalists left the heavy lifting to James Wilson in the former and James Madison in the latter, who could counter critics point for point while winning over undecided delegates by their visionary national outlook.

Late in 1787, Pennsylvania became the first state to take up the Constitution and the first big state to ratify it. On September 17, the very day that delegates to the Convention had signed the Constitution, those from Pennsylvania notified their state's assembly (meeting upstairs in the State House while the Convention used its normal chamber) that they "were ready to report" on the proposed new federal government.[5] The assembly agreed to hear their report on the eighteenth.

As senior delegate and state president, Franklin delivered the opening remarks, which began with the expression of his hope and belief that the new Constitution "will produce happy effects to this commonwealth, as well as to every other of the United States."[6] Sitting in seats that Convention delegates had occupied only a day earlier, members of the assembly then heard the Constitution and

Washington's transmittal letter read publicly for the first time. With this, the race was on to ratify, with the assembly scheduling Pennsylvania's ratifying convention to begin in late November and the legislatures of neighboring Delaware and New Jersey setting theirs to convene in early December.

Even though a federal Constitution was at stake, state constitutional issues inflamed the battle over ratification in Pennsylvania. Those issues already split state politics into two distinct parties that would morph into national ones. One party, rooted in rural central Pennsylvania, supported the existing state constitution, which concentrated power in a one-house legislature. Another party, led by Robert Morris and tied to Philadelphia's commercial interests, sought a balanced government with a two-house legislature and an independent executive. Of course, that very difference had animated the Constitutional Convention, with the latter view (favored by Washington) prevailing over the former. The reformers held a majority in the assembly and stacked the state's federal Convention delegation with its partisans.[7]

An artful politician, Franklin had friends in both state parties but Pennsylvania's other Convention delegates stood foursquare in the reform camp. Their thinking on state constitutional issues informed their contributions at the federal Convention. In contrast, backers of the old state constitution, excluded from the Convention, brought skeptical eyes to the proposed federal government, with its two-house legislature, strong executive, and elite Senate. The lack of a bill of rights heightened their concerns. In the election for delegates to the state ratifying convention, federalists secured two-thirds of the seats, which assured them of victory. They won all the seats in neighboring Delaware and New Jersey, which at the time circled within the economic orbit of Philadelphia and voted unanimously to ratify.

From start to finish, the federalist case for ratification in these three states relied on public trust in Washington and Franklin. "Remember, a WASHINGTON, a FRANKLIN, a MORRIS, with other illustrious, enlightened patriots composed it," an early federalist essay in Philadelphia's *Independent Gazetteer* said of the Constitution.[8] An antifederalist using the pseudonym "Centinel" countered by portraying Washington and Franklin as unwitting tools of partisan interests led by Robert Morris. "I would be very far from insinuating that the two illustrious personages alluded to, have not the welfare of their country at heart," he wrote, "but that the unsuspecting goodness and zeal of the one, has been imposed on; . . . and that the weakness and indecision attendant on old age, has been practiced on in the other."[9]

The federalist response came fast and furious. Within days of Centinel's blast, a letter to the *Gazetteer* denounced the antifederalists: "They do not reason, but abuse—General *Washington*, they (in effect) say, is a dupe, and Doctor *Franklin*, an old fool—vide the *Centinel*."[10] Many like it appeared.[11] "However respectable their names may be," one Philadelphia federalist essayist said of the antifederalists, "they cannot certainly be placed in competition with those of a Washington, a Livingston, a Franklin."[12]

In Washington's case, it mattered to many not only that he supported the Constitution but also that he would lead the government under it. "Should the Idea prevail that you would not accept the Presidency," Gouverneur Morris warned Washington about the Constitution's chances in the middle states, "it would prove fatal in many Parts."[13] Washington complied by modestly discounting the prospect rather than flatly dismissing it.

Having a solid majority, federalists at Pennsylvania's convention pushed through ratification without giving opponents an opportunity to offer amendments. This provoked a bitter reaction. Soon after the convention, a mob of unreconstructed antifederalists disrupted a federalist victory rally in Carlisle and burned James

Wilson in effigy.[14] Dissenters then published their amendments, circulated petitions asking the assembly to rescind ratification, and began working with antifederalists in other states to defeat the Constitution.

Word of these developments reached Washington early in 1788 and persuaded him that victory alone was not enough. He would have to rule these people, and he knew from the Revolutionary era that a disaffected minority could disrupt public order. When the local militia from nearby counties freed the imprisoned rioters in Carlisle to prevent their prosecution, the lesson should have become clear to all, yet some federalists wanted the rioters punished as an example.

Washington knew better. For the government to function, he reasoned, antifederalists would need to accept the ratification process as fundamentally fair. After having initially hailed the results in Pennsylvania despite the strong-arm tactics, Washington changed his tune. Offering advice to federalists in Massachusetts, where patriot warhorse Samuel Adams and Convention dissenter Elbridge Gerry were raising objections, Washington cautioned conciliation. "The business of the Convention should be conducted with moderation, candor & fairness," he advised Bay State federalists, "for altho' as you justly observe, the friends of the New system may bear down the opposition, yet they would never be able, by precipitate or violent measures, to sooth and reconcile their minds to the exercise of the Government." This reconciliation, Washington stressed, "is a matter that ought as much as possible to be kept in view."[15] Winning at all costs would not serve the public interest, he concluded. Whatever he thought of antifederalists, and various private letters betrayed his antipathy toward them, Washington knew better than to show it. He wanted to form an American nation by uniting its people, not dividing them.

WITH THREE MIDDLE states having ratified the Constitution by the end of 1787 but opposition rising elsewhere, the Massachusetts Convention played out as the pivotal act within the larger drama of federal ratification. It featured plot twists, intrigue, and a cliffhanger ending. From Mount Vernon, Washington followed the story in newspapers and private letters. At first, the prospects in Massachusetts looked rosy, with Washington receiving ever more upbeat reports over the course of the fall as federalists swept the Boston-area seats.[16] Estimating that nine-tenths of the people favored ratification, Boston's archfederalist *Massachusetts Centinel* asked in October, "Let it but appear that a HANCOCK, a WASHINGTON, and a FRANKLIN approve the new government, and who will not embrace it?" even though, at the time, federalists could not yet count on support from Governor John Hancock.[17]

Election returns from western Massachusetts and Maine in November suggested that the legacy of Shays's Rebellion was energizing the opposition.[18] Among antifederalists elected to the conclave, one former Constitutional Convention delegate lamented to Washington, there were "18 or 20 who were actually in Shay's army."[19] Now the *Centinel* asked, "Are the gentlemen who have withheld their assent to the Federal Constitution, superiour to Washington or Franklin, either in abilities or patriotism—men whose names, born on the wings of fame, are known throughout the world?"[20] Indeed, pro-ratification forces so often invoked Washington's support for the Constitution that the *Massachusetts Gazette* soon observed, "The *Federalists* should be distinguished hereafter by the name of *Washingtonians*, and the *Antifederalists*, by the name of *Shayites*."[21] The sides were closely matched. With a foot in each camp and influence matching his legendary ego, Hancock—who would serve as the convention's president—still had not tipped his hand.

When the convention opened on January 9, 1788, Hancock was

bedridden with gout. Uncertain if they had the votes to prevail, federalists opted to stall by agreeing to discuss the Constitution clause by clause before voting. This approach comported with Washington's advice. With practiced speakers on their side, including three returning delegates from the Constitutional Convention, federalists hoped to win over some undecided delegates, mollify the opposition, and play for time. They also received welcome help from Washington when, in the midst of the deliberations, a strident letter from him endorsing the Constitution appeared in nine Massachusetts newspapers. "There is *no alternative* between the *adoption* of it and *anarchy*," he declared in much underlined prose. "I am fully persuaded that it is the *best that can be obtained* at this time . . . and that *it* or *disunion* is before us."[22]

Federalists got all the delay they wanted. The discussion of Article I alone consumed two weeks as members wrangled over the authority of Congress, particularly its expansive taxing power. "In giving this power we give up everything," one antifederalist thundered. It is "as much power as was ever given to a despotic prince," another added. Federalists replied that since Congress would represent the people, any authority it held would serve the public.[23] Shaysites had objected to the concentration of power in the urban, moneyed elite under the state's 1780 constitution and feared a similar result under the new federal one. Many demanded amendments prior to ratification, even though no means existed to add those amendments without a second convention. By the last week of January, everyone knew the vote would be close, yet letters to Washington from leading Boston federalists sounded oddly confident—even smug.[24]

Their confidence rested on a secret deal with Hancock who, in exchange for support from federalists in the next governor's race, agreed to rise from his sickbed, endorse ratification, and propose postratification amendments to the Constitution.[25] The amendments would not be added "as a condition of our assent & ratification," a letter to Washington explained, "but as the Opinion of the Convention subjoined to their ratification."[26] In short, the convention would recommend that the first Congress consider them.

Federalists attached little importance to these "recommendatory" amendments, as Washington termed them, beyond their role in bringing aboard Hancock and perhaps winning over a few moderates.[27] Letters to and from Washington referred to them in dismissive terms.[28] Taken together, however, they offered outlines for the eventual bill of rights. Hancock deserves credit for advancing this key contribution to American democracy. He played his part to perfection by making a grand entrance at the convention swaddled in sick clothes and borne on a daybed just in time to carry the Constitution with his compromise proposal for recommendatory amendments. Wealthy beyond the dreams of most Americans of his day, Hancock always had a flair for the dramatic but never displayed it with greater effect than on this occasion.[29]

The margin was close—a 10-vote swing in the 370-member convention would have defeated the Constitution. Yet it passed by a whisker on February 6. Word of the results spawned the first widespread celebrations for ratification throughout the commonwealth of Massachusetts, which then included the district of Maine. In Berwick on February 11, the first of thirteen public toasts was raised to Washington; in Boston on the eighth, the second of thirteen; in Westminster on March 3, the fourth. The happenstance of these festivities coinciding or nearly coinciding with Washington's birthday on February 11 (under the old calendar, which many used to celebrate birthdays of persons born before the shift to the new calendar in 1752) likely played a role in the notice given to the Virginian, yet it was surely heartfelt. Boston's *Independent Chronicle* reported about the celebrations in Berwick, "The joy of the evening was heightened, by the pleasing reflection, that it was no less than the anniversary of that auspicious day that gave America, the most distinguished character in the world."[30] When the news reached New York, the last of thirteen toasts at a celebratory dinner attended by the mayor and members of Congress looked forward and back: to "General Washington—may his Wisdom and Virtue preside in the Councils of his Country."[31]

The victory cheered Washington as well, particularly how it was won. "The minority," a leading Boston federalist informed Washington on the day of the vote, "publickly declare that the Discussion has been fair & candid, and that the majority having decided in favor of the constitution, they will devote their Lives & Fortunes to support the Government."[32] Washington welcomed this result. "Happy, I am, to see the favorable decision of your Convention," he wrote back. "The candid, and open behaviour of the minority, is noble and commendable."[33]

Given the closeness of the final vote in Massachusetts, Washington's endorsement likely played a key role. The fact that the Constitution "comes authenticated" by Washington, one member told the state convention, "is a reason why we should examine it with care."[34] Just such careful examination led Massachusetts to become the second of four indispensable big states to ratify. Four smaller states ratified during the first five months of 1788, with the last of these, South Carolina, following the Massachusetts model of recommending amendments.[35]

These ratifications left the Constitution only one state shy of the needed nine by June, when New Hampshire and the two remaining big states, Virginia and New York, began their conventions. New Hampshire had tried and failed to ratify already, however, and antifederalists held the upper hand in New York. Only Virginia seemed in play, but far from certain.

Franklin reentered the fray at this point with a satirical letter to the nation's leading federalist newspaper comparing resistance to the Constitution with opposition to the Ten Commandments by the ancient Hebrews who received them from God by way of Moses—a story most Americans then knew. "I may not be understood to infer, that our general Convention was divinely inspired when it form'd the new federal Constitution, merely because that Constitution has been unreasonably and vehemently opposed," Franklin observed, "yet I must own I have so much Faith in the general Government of the World by PROVIDENCE, that I can hardly conceive a Transaction of such momentous Importance . . . should be suffered to pass without being in some degree influenc'd, guided and governed by" it.[36]

Believing that Virginia would decide the issue, Washington did all in his power to win its assent.[37] Ratification hung in the balance.

WARTIME EXPERIENCES HAD MADE WASHINGTON an American but he always remained a Virginian. In addition to being his home, Virginia was America's largest, wealthiest, most populous, and longest-settled state. Washington could not imagine a United States without it and worked hard to keep it in the fold.[38] On September 24, 1787, two days after returning to Mount Vernon from the Convention, he had sent copies of the Constitution to three former Virginia governors along with identical letters urging them to support it. Ratification, he wrote to the states-rights-minded Patrick Henry, Benjamin Harrison, and Thomas Nelson, "is in my opinion desirable."[39] With returning convention delegate George Mason breathing fire and Henry inclined to join him, Washington feared the Constitution faced strong headwinds in Virginia.[40] Over the ensuing months, he regularly hosted Virginia federalists at Mount Vernon and lobbied current governor Edmund Randolph and former governor Thomas Jefferson in letters designed to bring or keep them on board, leading to rumors that he promised them

posts in the first presidential administration. Both later joined
Washington's first cabinet.

Some 170 members strong, Virginia's ratifying convention
opened on June 2 in Richmond, the first contested convention
not held in a federalist-friendly port city. After having tried to
win him over early, Washington worried more about Henry's
opposition than about that of any other Virginian, to the point
of fearing that the former governor might seek to lead the state
into a separate southern confederacy with himself at the helm.[41]
While not a disciplined debater, Henry had a gift for stirring au-
diences with impassioned speeches that played on emotion.[42] He
employed this approach at the convention to transform what was
supposed to be a clause-by-clause consideration of the Constitution into a free-for-all in which federalists scrambled to refute his
scattered charges.

Henry launched his assault on the first day of substantive debate
and did not spare even Washington. Representing the country as
being in "universal tranquility" prior to the Convention, Henry
demanded to know why Virginia's delegates—particularly
Washington—had proposed replacing the confederation of states
with a consolidated national government. "Who authorized
them to speak the language of, *We, the People,* instead of *We, the
States*?" he asked. "I would demand the cause of their conduct,"
he said, "even from that illustrious man who saved us by his
valor." Liberty, Henry charged, was at stake. "This proposal of
altering our Federal Government is of a most alarming nature,"
he declared, "for instead of securing your rights you may lose
them forever."[43]

The public challenge to Washington drew gasps from federalists and brought Randolph to his feet for a two-hour-long oration.
To this point, Henry still saw Randolph as an ally and expected
his support.[44] Instead, the governor defended the Constitutional

Convention. "The gentleman," he said of Henry, "inquires, why we assumed the language of 'We, the people,' I ask why not? The Government is for the people; and the misfortune was, that the people had no agency in the Government before." With "the terror of impending anarchy" and no hope of saving the confederation, Randolph asked in defense of Washington, "Would it not have been treason to return without proposing some scheme to relieve their distressed country?" He withheld his signature from the Constitution thinking that the states would insist on amendments prior to ratification, Randolph explained, but the actions of eight states now showed otherwise.[45]

As he would throughout the convention, Madison promptly reported to Washington. "The Governor has declared the day of previous amendments past, and thrown himself fully into the federal scale," Madison wrote in a same-day letter to Mount Vernon. "The federalists are a good deal elated by the existing prospect." Elated but cautious, Madison added, because the delegates from Virginia's western districts seemed uniformly hostile—and given the close divide, even a small contingent could prove decisive. "Every piece of address is going on privately to work on the local interests & prejudices of that & other quarters," Madison assured Washington.[46]

Between letters and newspaper reports, Washington received a blow-by-blow account of the convention. For two weeks, it did not proceed in a systematic fashion but instead followed Henry's lead as he discharged random "bolts," as Henry Lee called them, raising all the familiar antifederalist complaints. The president would become a king, Henry charged, and the federal government, the people's master. Property could be taxed into oblivion, he warned, and without a bill of rights, American liberty would be lost.

Madison and his allies responded point by point. The convention picked up speed only after delegates began reviewing the document clause by clause.[47] "Our progress is slow and every advantage is taken of the delay, to work on the local prejudices of particular setts of members," Madison explained to Washington in mid-June.[48] As the vote neared, Madison advised Washington, "We calculate on a majority, but a bare one."[49]

Although he never left Mount Vernon, Washington was a virtual presence within the assembly room at Richmond. His role in writing the Constitution and the prospect of his becoming president once again made all the difference. Alluding to him near the convention's end, one antifederalist delegate complained, "Were it not for one great character in America, so many men would not be for this Government."[50] When the convention ended, another— the future president James Monroe—wrote about Washington, "His influence carried this government."[51] Washington's assumed role as president made what they saw as a fatally flawed system appear attractive to others.

ON JUNE 27, 1788, the evening stagecoach brought the news to Alexandria that Virginia had ratified the Constitution two days earlier. It had passed by a ten-vote margin. Using the Massachusetts model of ratifying with recommendatory amendments carried the day. Townspeople descended on Mount Vernon to invite Washington to festivities scheduled for the next day.

Before dawn, an express rider arrived in Alexandria with word that New Hampshire had ratified on the twenty-first, making it, not Virginia, the ninth state to approve the Constitution. "Thus the Citizens of Alexandria, When convened, constituted the first public company in America, which had the pleasure of pouring libation to the prosperity of the ten States that had actually adopted the general government," Washington noted. "The Cannon roared, and the Town was illuminated." The Constitution would take effect whether or not the remaining three states joined, but Washington felt sure that they would rather than be left out altogether. "The point in debate has, at least, shifted its ground from policy to expediency," he astutely observed.[52]

Federalists in cities and towns across the country erupted in similar celebrations, typically hailing ratification as something between a rebirth of nationhood and the culmination of the revolutionary struggle. America was free, they cheered, and secure. Yet accounts and images of these festivities show mostly men participating, and virtually no Blacks or Natives. In Alexandria, for example, only white men attended the celebratory banquet, with women joining the subsequent dance. For Washington, that meant leaving Martha behind at Mount Vernon and attending the festivities with his nephew George and his longtime aide David Humphreys.[53] Clearly, the American project remained unfinished.

For their part, Philadelphia federalists celebrated ratification with a grand parade featuring Franklin on July 4—the twelfth anniversary of the signing of the great declaration that he had coauthored. Military corps, officeholders, and bands marched up Market Street near Franklin's house. Units representing some forty male vocations followed in line. Carpenters on one carriage worked on a structure titled "GRAND FEDERAL EDIFICE"; printers on another struck off broadsheets bearing Franklin's words. The pageant presented a cross section of the local economy except for women's work and the labor of slaves. No female appeared in the list of participants and, even in Quaker Philadelphia, the only Native American represented was portrayed by a white man smoking a peace pipe with another white man on a horse-drawn carriage. Banners named the four great Revolutionary era events in which Franklin had a central role: Independence, Alliance with France, Treaty of Peace, and Constitution.[54] The flags of allied countries did include one of Muslim Morocco, but, unlike the others, it was not carried by an immigrant from that place. As these pageants suggest, ratification came at a price.

A TRIAL LAWYER renowned for winning cases by appealing to jurors' emotions, Patrick Henry had raised one notable objection at Virginia's ratifying convention that antifederalists in Pennsylvania and Massachusetts did not. *"They'll free your niggers!"* he reportedly exclaimed to punctuate his harangue against the expansion of federal powers under the Constitution.[55] "We ought to lament and deplore the necessity of holding our fellow-man in bondage," Henry conceded, "but is it practicable by any human means to liberate them?" Emancipation would ruin the south, slaveholders argued. Henry himself owned some seventy slaves, and never freed one. But because "slavery is detested," he declared, a fortified federal government controlled by nonslaveholding northerners would "clearly and certainly" end it. "Among the ten thousand implied powers which they may assume," Henry thundered about Congress (using a plural pronoun for that composite body), "they may, if we be engaged in war, liberate every one of your slaves."[56]

Even without invoking war powers, Henry said of Congress under the Constitution, "Have they not power to provide for the general defence and welfare? May they not think that these call for the abolition of slavery? May they not pronounce all slaves free, and will they not be warranted by that power?"[57] At the very least, he declared, Congress will burden slave owners with ruinous taxes "so as to compel the owners to emancipate their slaves rather than pay the tax."[58] To white Virginians, the abolition of slavery represented the ultimate threat and free Blacks embodied a terrifying peril. Federalists had to answer.

"I was struck with surprise when I heard him express himself alarmed with respect to the emancipation of slaves," Madison shot back about Henry. "There is no power to warrant it in" the Constitution. No American Congress would ever weaken the nation by abolishing slavery and thereby "alienate the affections of, five-thirteenth of the Union," Madison insisted. "I believe such an idea never entered into any American breast, nor do I believe it ever will, unless it will enter into the heads of those Gentlemen who substitute unsupported suspicions to reasons."[59] This was a conspiracy theory hatched in the minds of southern antifederalists to discredit the Constitution, he asserted. "If any Gentleman be terrified by this apprehension, let him read the system. I ask, and I will ask again and again, till I be answered (not by declamation) where is the part [of the Constitution] that has a tendency to the abolition of slavery?" Randolph added. "Were it right here to mention what passed in the Convention on the occasion, I might tell you that the Southern States, even South Carolina herself, conceived this property secure."[60]

To calm the delegates, Madison and Randolph reviewed the various provisions inserted in the Constitution to protect and propagate slavery. The Three-Fifths Compromise augmented slave-state representation in Congress and the Electoral College while limiting the impact of a head tax on slaves, they noted. One clause abetted the return of fugitive slaves, another barred Congress from outlawing the slave trade until 1808, and a third limited the import tax on slaves.

The ultimate protection for slavery lay in the structure of government under the Constitution, Madison and Randolph added. Congress holds enumerated powers, Madison noted, and "no power is given to the General Government to interpose with respect to the property in slaves now held by the States."[61] Randolph demanded of Henry, "Point out the clause whereby this formidable power of emancipation is inserted."[62] Concerns stemming from assaults on slavery by several northern delegates may explain why Randolph, midway through the Convention, orchestrated the substitution of the Virginia Plan's broad assertion of federal authority "in all cases to which the separate States are incompetent" with an enumerated list of federal powers.[63] And it may help account for the countermove by James Wilson, a critic of slavery, to add the sweeping necessary and proper clause to those listed powers. Slavery shaped the Constitution.

Henry remained adamant. "He asked me where was the power of emancipating slaves. I say it will be implied," Henry said of Madison and his Constitution. "He says that I am unfair and uncandid in my deduction, that they can emancipate our slaves, though the word emancipation be not mentioned in it. They can exercise power by implication."[64] Madison may not have satisfied Henry, but he stilled the waters enough for the Constitution to win the convention's approval, and with it to become the law of the land, keeping Washington on track to the presidency. Yet while Madison won the battle for ratification, Henry had boxed him into an absolutist position on federal power. The now inflexible stance of Madison on slavery and the Constitution put him, and Washington, on a collision course with Franklin.

EVENTS AT THE CONSTITUTIONAL CONVENTION belie Madison's pledge in Richmond that he believed the idea of the new federal government's abolishing slavery "never entered into any American breast." He knew otherwise. Certainly the slaves themselves dreamed of freedom and often risked their lives to secure it. In 1780, Pennsylvania (where the Convention was held) became the first state to end slavery, doing so gradually by barring any new slaves born in or brought into the state. In 1784, a Quaker-led Philadelphia antislavery association reorganized itself into a

more ambitious Pennsylvania Society for Promoting the Abolition of Slavery (or Pennsylvania Abolition Society), with a mission to free slaves throughout the United States. In 1787, before
the Convention began, Franklin became the society's president.
Society members urged him to raise the issue of manumission at
the Convention but, according to a society vice president, Franklin decided that "it would be a very improper season & place" to
do so because of the opposition to constitutional reform it would
generate in slave states.[65] The most effective way to abolish slavery in southern states, Franklin believed, was to bring them into a
fortified federal union, and then (just as Henry warned) have the
government work toward that end.

At the Convention, while Franklin remained quiet, other northern delegates railed against slavery. "It was a nefarious institution,"
Pennsylvania's Gouverneur Morris said. "And What is the proposed
compensation to the Northern States for a sacrifice of every principle of right, of every impulse of humanity [in accepting it under
the Constitution]. They are to bind themselves to march their militia for the defense of S. States; for their defense agst those very
slaves of whom they complain."[66] Slaves would revolt, Morris assumed, and the nation would now have to suppress them.

Rufus King of Massachusetts expanded on Morris's point in a
manner that echoed an old but well-known quip by Franklin. Noting that the main object of the union was defense against foreign
invasions and internal insurrections, King challenged constitutional protections for the slave trade. "Shall all the States be bound
to defend each," he asked his fellow delegates, "& shall each be at
liberty to introduce a weakness which will render defence more
difficult?"[67] Slavery created the ever present risk of slave revolts,
every delegate knew, yet the Constitution barred ending the slave
trade until 1808, limited the taxation of slaves, and forbade taxing

exports, which disproportionately came from slave plantations. King thought the heightened risk of insurrections justified more, not less, taxes from the slave states.

Franklin had raised a similar issue at the Continental Congress in 1776, when he defended apportioning taxes to the states by population, including slaves. "Our Slaves are our Property," one South Carolina congressman retorted, "why should they be taxed more than . . . Sheep?" Sheep differ from slaves, Franklin replied, "Sheep will never make any Insurrections."[68] In that same month, Franklin defended passages in the draft Declaration of Independence condemning the slave trade—passages deleted at the demand of southern members. Franklin, Madison knew, harbored the dream of abolition in his breast.

If the debate in Philadelphia did not alert Madison to the abolitionist designs of some delegates, then the deliberations at earlier ratifying conventions surely did—deliberations he followed closely. As a means to undermine support for the Constitution in Pennsylvania, Massachusetts, and other northern states where slavery was detested, some antifederalists railed against constitutional protections for slavery. Federalists countered by arguing that, to the contrary, the constitutional union offered the best possible means of eventually eradicating slavery from all the states. No delegate carried more weight on this score that James Wilson, who with Madison and Morris stood out as one of the key architects of the Constitution.

Six months prior to the Virginia ratifying convention, when one antifederalist abolitionist at Pennsylvania's convention damned the clause barring Congress from restricting the slave trade until 1808, Wilson had a ready response. "Under the present Confederation, the states may admit the importation of slaves as long as they please," he said, "but by this Article, after the year 1808, the Congress will have power to prohibit such importation, notwithstanding the disposition of any state to the contrary. I consider this as laying the foundation for banishing slavery out of this country." Wilson wished that slavery could end sooner, he added, "but from this I think there is reason to hope that yet a few years and it will be prohibited altogether; and in the meantime, the new states which are to be formed will be under the control of Congress in this particular; and slavery will never be introduced amongst them."[69] What to Henry was a dark threat to slavery, to Wilson was the bright promise of emancipation, and Franklin instinctively looked on the bright side.

FRANKLIN HAD NOT ALWAYS OPPOSED SLAVEHOLDING. A global institution dating from time immemorial, slavery was a fact of life during Franklin's youth, even though his own escape from an indentured apprenticeship—a form of involuntary servitude—at age seventeen may have sensitized him to the plight of slaves. He later suggested as much.[70] Yet in 1723, when Franklin arrived in Philadelphia to seek his fortune (as he put it) "from the Poverty and Obscurity in which [he] was born and bred," even Quakers owned slaves.[71] As a young printer, he published ads for slaves in his newspaper; as a middle-aged publisher, he owned one or more house slaves at any given time—perhaps seven in all. Franklin's wife, Deborah, sent their young slaves to a mission school for enslaved children. Franklin was so impressed with the aptitude of students at this school that he became an advocate of Black education. "Their Apprehension seems as quick, their Memory as strong, and their Docility in every Respect equal to that of white Children," Franklin commented on these students to a school official, attributing his prior contrary views to "my Prejudices."[72] In 1757, six years after publishing observations decrying the effects of slavery on society, Franklin changed his will to provide freedom for his slaves upon his death.[73] Ten years later, he had none.

Franklin's thinking on slavery had evolved in the half century from 1720 to 1770, so that he was no longer comfortable owning slaves, but he was not yet an abolitionist. That took longer. As early as 1729, Franklin began publishing the writings of Quaker abolitionists—decades before their views became church dogma.[74] A journalist at heart, Franklin believed in presenting all sides. "Printers are educated in the Belief," he wrote in 1731, "that when Truth and Error have fair Play, the former is always an overmatch for the latter: Hence they chearfully serve all contending Writers that pay them well, without regarding on which side they are of the Question in Dispute."[75] But Franklin did more than publish the views of abolitionists. He listened to them, particularly to Benjamin Lay, who lived like a hermit in a cave outside Philadelphia and persisted in shocking Quakers into confronting the sinfulness of owning people. Lay went so far as to kidnap children briefly from smug Quaker slaveholders, in order to show them how it felt to suffer abduction. By 1758, Deborah had hung Lay's portrait in the Franklin home, and her husband was wondering where she got it.[76]

Spiritual pleas alone could not persuade a rationalist like Franklin. The self-evident sin of the slave trade—snatching people from their homes in Africa and transporting them in hellish holds into New World bondage—turned him against that practice by 1760, when he began attacking it in print. Only during his residence in France during the American Revolution, however, did the Enlightenment era arguments of such philosophes as Nicolas de Condorcet finally convince Franklin that slavery itself—not just the slave trade—should end. Returning in 1785 to a Philadelphia almost cleansed of slavery, he began, for the first time, speaking openly and unequivocally against the institution. Two years later, he accepted the presidency of the Pennsylvania Abolition Society—an office he held for the rest of his life. From this perch, he struck.

DURING THE FALL OF 1788, the first federal elections for Congress and president coincided with the end of Franklin's third term as Pennsylvania's chief executive. Term limits barred him from serving again and, due to deteriorating health, Franklin welcomed retirement. The post would soon change to that of governor under a new state constitution. Franklin knew it was time for him to slow down. By then a firm believer in providence guiding America toward a noble destiny as the world's first continental republic and a global beacon of liberty, Franklin embraced the transition from sovereign states to a federal union.[77]

Franklin had planned to retire in 1787, at the end of his second term, but acceded to the unanimous vote of the state's assembly to continue. At that earlier time, he had hoped to take one last trip to Boston, seeing his sister and revisiting sites of his youth. A bad fall in January 1788 had severely limited his mobility and aggravated his kidney stone, keeping him from meetings of the executive council in the State House for months and barring future travel. "It certainly would," Franklin wrote in mid-1788, "be a very great Pleasure to me, if I would once again visit my Native Town, and walk over the Grounds I used to frequent . . . and where I might find some of my old Acquaintance." He could no longer tolerate the long journey or bear walking on Boston's cobbled streets, however, and added with a touch of heartache, "I should find very few of my old friends living."[78] Although Franklin's mind remained agile and his wit sharp, he never again ventured far from his happy home on Market Street, where virtually his entire immediate family lived with him.

Even as Franklin was retiring from public life, Washington

reluctantly prepared to resume his. Over the autumn of 1788 and winter of 1789, the first federal elections went pretty much as Franklin and other astute observers expected, at least with respect to the presidency. Under the new Constitution, each state could determine its own means for choosing presidential electors, and all scrambled to do so. Four middle states opted for voters to elect them either by districts or at large. Five states entrusted the task to their state legislatures but in one of those states, New York, the process became so bogged down in partisan wrangling that it failed to choose any. Not yet having ratified the Constitution, North Carolina and Rhode Island did not participate in the elections. The two remaining states, Massachusetts and New Hampshire, used methods that mixed popular voting with legislative selection. The means did not matter. When the electors of the ten participating states met on February 4, 1789, all sixty-nine of them cast one of their two votes apiece for Washington. Their second votes scattered, but John Adams received enough of them to become vice president. Federalists aligned with Washington captured a majority of the seats in both houses of Congress even though, under Henry's leadership, the Virginia legislature sent two antifederalists to the Senate.

Although Washington knew by January that he would win the electoral contest, and by February that the count for him would likely be unanimous, the results could not become official until Congress convened and counted the votes. That took until early April. "On this day we went to business, and to my very great satisfaction I heard an unanimous vote of the electing States in favor of calling you to the honorable office of President," Virginia senator Richard Henry Lee wrote to Washington on April 6, 1789. "I am sure that the public happiness, which I know you have so much at heart, will be very insecure without your acceptance."[79] After receiving official notice of his selection, Washington left

Mount Vernon on April 16 on a ten-day carriage trip to New York for his inauguration. With formal receptions and cheering crowds at every town along the way, it became a grand federal parade honoring Washington and uniting the nation.

Reaching Franklin's Pennsylvania on the twentieth, Washington was greeted by the state's new president at the border and—with other officials, two cavalry units, a detachment of artillery, and a body of light infantry—escorted through a festooned ceremonial arch into Philadelphia, which exploded upon his arrival. Cannons fired and bells rang through the day; fireworks lit the night's sky. "The number of spectators who filled the doors, windows, and streets, which he passed, was greater than on any other occasion," a newspaper noted. "All classes and descriptions of citizens discovered . . . the most undisguised attachment and unbounded zeal for their dear chief."[80] A banquet followed.

Infirm to the point of being virtually bedridden, Franklin could not attend the public events for Washington but received him at his home. No record remains of what was said, even though both surely knew it would be their last meeting. After spending the night at Robert Morris's mansion, Washington proceeded on to New York City, where Congress sat. There, on April 30, before thousands of spectators spread across the streets and spaces below, Washington took the oath of office on an outside balcony of the city's newly renovated Federal Hall on Wall Street.

After taking the oath in public, Washington returned to the Senate Chambers and, in a voice that the assembled dignitaries strained to hear, delivered an inaugural address setting out his vision for the new republic. "The preservation of the sacred fire of liberty, and the destiny of the republican model of Government," he said in lofty words that echoed Franklin's beliefs, "are justly considered as deeply, perhaps as finally, staked, on the experiment entrusted to the hands of the American people."

Looking heavenward, Washington closed with a prayer for the American people to exhibit "the enlarged views, the temperate consultations, and the wise measures, on which the success of this Government must depend."[81] For Washington, those measures would include Treasury Secretary Alexander Hamilton's ambitious plans of government-assisted economic development through protective tariffs, debt assumption, and a quasi-independent central bank. They did not include tackling such sectionally divisive and economically disruptive matters as the abolition of slavery or the slave trade. Here Franklin's view of America's lofty duty to preserve the sacred fire of liberty differed from Washington's notion of the temperate consultations needed for successful government.

SLAVERY WAS ONE MATTER that irreconcilably divided Franklin and Washington, just as it became the subject that, seventy years later, tore apart the states that they had worked so long to knit together. Coming from the south and knowing full well the issue's divisiveness, temporizing on slavery was nothing new for Washington. It was his practice.

Born into a slaveholding family, Washington owned more than one hundred slaves and controlled nearly two hundred more in his wife's dower estate by the time of the American Revolution. During and after that fight for liberty, critics of slavery, from Quaker abolitionists to his much loved military aide Lafayette, pleaded with Washington to denounce the institution publicly or at least set an example by freeing his own slaves. While he sometimes sympathized with such views in private, he always equivocated or delayed acting. Meanwhile, the number of his own slaves increased, he never freed any during his lifetime, and he pursued those who ran away. Dozens fled, including seventeen in 1781 when a British warship approached Mount Vernon with promises of freedom for slaves who betrayed their masters.[82] Those in human bondage who knew his private face never saw Washington as a liberator.

Historians disagree on whether Washington was an unusually harsh master. Certainly he clothed and housed his field slaves poorly, worked them hard from sunrise to sunset, and had them whipped for perceived infractions. Although he favored mixed-race slaves for personal tasks and kept those sired by his father-in-law, stepson, and other relatives in bondage at Mount Vernon, he appeared to abhor (and never practice) the acts that produced them—but Washington was not a lustful male, and did not have children by Martha either. Opportunity tempted him only in warfare and real estate. Franklin once observed that "almost every Slave [was] *from the nature of slavery* a thief."[83] Washington saw slaves as thieves as well but, lacking Franklin's empathy, blamed them rather than their bondage. He distrusted his slaves, and they distrusted him. In short, he treated them like slaves. They were his human chattel.

While serving as president after Philadelphia became the capital in December 1790, seven months after Franklin's death, Washington began rotating his house slaves back to Virginia. He did this to avoid a Pennsylvania law liberating slaves held within the state for more than six months. "I wish to have it accomplished under pretext that may deceive both them and the Public," he wrote to his personal secretary about the rotation. "This advice may be known to none but *yourself & Mrs. Washington*."[84] Deducing what was afoot, Washington's chef, Hercules, so heartily protested his fidelity that Washington exempted him from the policy. When the time came for him to return to Virginia, Hercules fled. Martha's favorite attendant, Ona Judge, did so too. By this time, Washington had signed the nation's first fugitive slave act into law and used every means it offered to retrieve his runaway bond servants. He had been in no mood to cooperate earlier in 1790, when Franklin lashed out against slavery. Giving Washington the benefit of the doubt, one could attribute his response to prioritizing the preservation of the union over justice to slaves—but little in his prior public acts merit such a concession. To the people, he appeared a stout defender of slavery. When Franklin's closest abolitionist friend had died in 1784, a eulogist declared, "I would rather be Anthony Benezet in that Coffin than George Washington, with all his fame."[85]

FRANKLIN DID NOT want to go to his coffin without one last swipe at slavery. Since taking the reins of the Pennsylvania Abolition Society in the spring of 1787, he had assumed an ever more prominent position criticizing the slave trade and urging emancipation. He pleaded with influential slaveholders such as Virginia governor Edmund Randolph, who was named the nation's first attorney general in 1789, to free their slaves, and he urged New Hampshire president John Langdon to discourage his state's merchant shippers from participating in the slave trade. Franklin had hosted both men at his home during the Constitutional Convention and felt free to press them on the issue. Slavery "is so evidently repugnant to the political principles and form of government lately adopted by the citizens of the United States," he wrote to Langdon, that it "cannot fail of delaying the employment of the blessings of peace and liberty by drawing down the displeasure of the great and impartial Ruler of the Universe upon our country."[86] Franklin believed that a republican people must exhibit civic virtue to prosper. Slavery threatened calamity.

Revered as much in Europe as America, Franklin corresponded with leading abolitionists across the Atlantic to coordinate a global campaign against slavery. English industrialist Josiah Wedgwood sent him literary sketches about freed Blacks. "While relief is given to so many," Wedgwood wrote, "the subject of freedom itself will be more canvassed and better understood."[87] To Lafayette, Franklin pleaded about slavery, "Nothing effectual will be done in this business untill France concurs in it."[88] In mid-1788, London classicist and social reformer Granville Sharp shared news that "[u]pward of a hundred petition having been presented to Parliament" against the slave trade, "the House of Commons pledged itself to take up the business."[89]

Perhaps this news from London inspired Franklin. His final assault on slavery took the form of a petition to Congress that he signed as president of the Pennsylvania Abolition Society. Declaring "that equal liberty was originally the Portion, and is still the Birthright of all Men," it called on members of Congress to "step to the very verge of the Powers vested in you, for discouraging every Species of Traffick in the Persons of our fellow Men." Paraphrasing the Constitution to say those powers included "promoting the Welfare and securing the blessings of liberty to the People of the United States," the petition asserted, "These blessings ought rightfully to be administered, without distinction of Color, to all descriptions of People."[90]

Franklin's petition reached the House of Representatives on February 12, 1790, one day after similar petitions arrived from two Quaker societies. At the time, Congress routinely referred citizens' petitions to committee for consideration. The Quaker petitions stirred an immediate uproar, with some southern members opposing their referral. The petitions urged Congress to oppose the slave trade and discourage slavery, which these members claimed was beyond its power. In an instructive precedent for originalist interpretation of the First Amendment, they added that, because the petitions came from churches instead of citizens, acting on them would constitute a wrongful establishment of religion.[91]

The vehemence of the opposition did not stem from these procedural concerns, however, but from the petitioners' known objectives. "Do these men expect a general emancipation of slaves
by law?" South Carolina's Thomas Tudor Tucker thundered.
"This would never be submitted to by the southern states without civil war." Even referring antislavery petitions to committee
risked disunion, Georgia's Abraham Baldwin warned. "The moment we go to jostle on that ground," he said, "we shall feel it
tremble under our feet." Yet by opposing referral, these members
ignited the debate that they sought to avert by stirring northern
members to rise in defense of the petitioners. "It was the cause of
humanity they had interested themselves in," Elbridge Gerry of
Massachusetts declared, "to wipe off the indelible stain which the
slave trade had brought upon all who were concerned in it." Then
let them read the Bible, James Jackson of Georgia replied, and
"they will find that slavery is not only allowed but commended."[92]
Already in 1790, all of the major antebellum arguments for and
against slavery sounded in Congress. "More masterly speeches
could not have been heard, upon a similar subject, in any part of
the world," one newspaper observed.[93]

Derailing the ongoing discussion of Hamilton's landmark financial plan for restructuring government debt and boosting the
national economy, the debate over referring the Quaker petitions
extended into the next day, when the petition from Franklin arrived and effectively forced Congress to refer all three to committee. Franklin's signature gave the lie to opponents' procedural
arguments, with Tucker vainly protesting that Franklin "ought
to have known the Constitution better."[94] Quietly referring the
petitions to committee was what James Madison wanted anyway. A shrewd political strategist then representing Virginia in
the House of Representatives, he knew better than anyone that,

while Congress could not constitutionally bar the slave trade until 1808, it could use its substantial powers to undermine slavery and that he could more effectively counter that effort by using (rather than resisting) legislative action. "The best way to proceed in this business is to commit the memorials without any debate," the slaveholding congressman advised his House colleagues.[95] Northern members wanted this result anyway and, with Madison on their side, they easily prevailed over divided southern opposition. And so the petitions went to committee. On this tinderbox issue, Madison sought to forestall action through process.

BY THE TIME he submitted his petition to Congress, Franklin knew he was dying. He had written his last letter to Washington on September 16, 1789—several months before the antislavery petition was sent. "For my own personal Ease, I should have died two Years ago; but tho' those Years have been spent in excruciating Pain, I am pleas'd that I have liv'd them, since they have brought me to see our present Situation," Franklin wrote of the period that covered the Constitution's ratification and new government's founding. "I am now finishing my 84th and probably with it my Career in this Life; but . . . if I retain any Memory of what has pass'd here, I shall with it retain the Esteem, Respect, and Affection with which I have long been, my dear Friend, Yours most sincerely, B. Franklin."[96]

Washington replied a week later, expressing his prayer "that your existence might close with as much ease to yourself, as its continuance has been beneficial to our Country and useful to Mankind." Echoing back Franklin's esteem, Washington closed with the salutation, "So long as I retain my Memory, you will be thought on with respect, veneration and affection by Dear Sir

Your sincere friend, G. Washington."[97] A fitting end to a three-decade-long relationship, it preceded their final letterless encounter over slavery.

Franklin suffered from the stone and gout. To combat them, he wrote shortly before his letter to Washington, "I have been obligated to have recourse to Opium, which indeed has afforded me some Ease from time to time but then it has taken away my Appetite and so impeded my Digestion that I am become totally emaciated and little remains of me but a Skeleton covered with a Skin."[98] To his satisfaction, this relief allowed him to resume writing his autobiography, which became the best-selling American book of the next century and a worldwide classic of its genre. By November, Franklin complained that he no longer had "any faith in remedies," and simply sought palliative relief to "make life at least tolerable." Unable to sit, he added, "I now make use of the hands of one of my grandsons, dictating to him from my bed."[99] Visiting Franklin in early March during the break in the congressional battle over the antislavery petitions, Thomas Jefferson found him "much emaciated, but in good spirits," and animated "almost too much for his strength." Franklin's mind remained sharp.[100]

AFTER THE SPECIAL COMMITTEE charged with considering the antislavery petitions finished its work in early March, the full House took up its report on the sixteenth. Congress had named one member from each state to the committee, but since no one from South Carolina and Georgia would serve and North Carolina was not yet a state, the group had a decidedly northern tilt. Without making any policy recommendations, the committee simply reported seven enumerated propositions of what, under the Constitution, Congress could do about slavery. Construing the Constitution narrowly, the first three propositions stated that Congress could not prohibit the importation of slaves until 1808, could never emancipate them, and could not interfere with the internal regulation of slavery in the states. The next four, however, said that Congress could impose a ten-dollar tariff on imported slaves, could regulate conditions on slave ships, could bar foreign carriers from the U.S. slave trade, and could exercise these powers humanely.

The committee's report outraged House members from Georgia and South Carolina. They moved to strike it and substitute one rejecting the petitions as "unconstitutional, and tending to injure some of the states."[101] The debate over this motion consumed a day and produced the most strident defense of slavery heard during the entire episode, most of it coming in one long harangue by Georgia's James Jackson.

Jackson denounced the Quakers as self-righteous zealots who (as pacifists) did nothing to win the Revolution. Unable to level a similar charge against Franklin, who everyone knew had played a critical role in the war effort, Jackson dismissed him as senile. Painting a grossly inaccurate picture of horrific conditions in Africa and an absurdly happy one of plantation life, he professed that Africans "would be better imported here by millions than stay in their own country." Similar reasoning applied against returning them to Africa, where Jackson said they would be reenslaved by African despots. Pointedly but incorrectly asserting that white people would never intermix with Blacks, Jackson concluded that slavery was the best option for all concerned and perfectly compatible with the commands of "Jesus Christ, who allowed it in his day and his apostles after him." Slavery being established in the south and essential to the regional economy, he declared, "Congress cannot interfere without endangering the whole system of government: that excellent constitution which we have so happily effected."[102] As members rose to refute Jackson's various points, the chair ruled the motion to strike the committee's report out of order, and the House settled down to address each proposition separately in the deliberate fashion that Madison wanted from the outset.

The legislative process took five more days, but one by one by narrow margins—some by but a single vote with odd-fellow coalitions of slave-owning southerners and pragmatic northerners—the House pared the committee's seven propositions to three. These declared Congress ineligible to ban the importation of slaves until 1808 or to interfere with slavery in the states but free to regulate foreign sales of slaves and conditions on slave ships.[103] Not all agreed, of course. Some members insisted that Congress could end the slave trade earlier and emancipate slaves at any point, but a majority shepherded on the floor by arguments and motions made by Madison ruled otherwise.

With the House report as precedent, Congress barred further debate on the slave trade and any discussion of emancipation. To seal the arrangement, Madison obtained consent to have the amended report memorialized in the House *Journal,* where it could stand as a guide to future action. Congressional power over slavery "has been so fully discussed," the *Journal* stated, "it cannot be supposed that gentlemen will go over the same ground again."[104] On the most divisive issue in antebellum American politics, silence would reign in the halls of Congress.

Philosophically opposed to slavery but seeing no way to end it, Madison (much like Washington) shuddered when politicians like Jackson defended it as a positive good or when reformers like the Quakers damned it as an intolerable evil. Either approach invited disunion. Madison simply wanted the issue to go away. At the Constitutional Convention, he had warned that the fundamental fault line dividing America was not between small and large states, or rich and poor ones, but between states with slaves and states without them. The debate over these petitions redoubled his concerns. As a member of Congress and later as secretary of state and president, Madison sought to paper over the problem of slavery and, by doing so, he perpetuated it.

Franklin and the Quakers submitted their antislavery petitions before Washington had fully formed his cabinet or gained his footing in dealing with Congress. At that time, Madison served as the president's chief advisor and legislative spokesperson. Hamilton eventually assumed the former role and Madison lost the latter one, but in 1790, he was known as Washington's prime minister. As such, he surely conferred with Washington on the petitions, which preoccupied Congress for parts of February and March. Certainly, Washington approved of Madison's handling of them, saying so privately.[105]

Washington in-law David Stuart, a Virginia politician, complained about those petitions in a mid-March letter to the president in which he warned about the attitudes of Virginians toward northeastern (or "Eastern") states, "A spirit of jealousy which may become dangerous to the Union . . . seems to be growing fast among us."[106] Washington replied to Stuart shortly after Madison had defanged the petitions with his House report. "That there is diversity of interests in the Union none has denied," Washington conceded. "A spirit of accomodation was the basis of the present Constitution; can it be expected then that the Southern or the Eastern part of the Empire will succeed in all their Measures? certainly not. But . . . if the Eastern & Northern States are dangerous *in Union,* will they be less so *in seperation?*" The south, he warned, would "most unquestionably, be the weaker party." Turning to Madison's resolution of the matter of the petitions, Washington smugly added, "The Memorial of the Quakers (& a very malapropos one it was) has at length been put to sleep, from which it is not [likely] it will awake before the year 1808."[107] On this issue, he too favored silence.

FRANKLIN DID NOT. Jackson's defense of slavery stirred him to write one last sharp satire—perhaps his best. Diminished to bedridden skin and bones, the mind within his emaciated, pain-racked body was as vigorous (and wit as slashing) as ever. "Reading last night in your excellent paper the speech of Mr. Jackson in Congress," Franklin wrote in a March 23 letter to the *Federal Gazette* under the pseudonym Historicus, "put me in mind of a similar one made about one hundred years since, by Sidi Mehemet Ibrahim, a member of the Divan of Algiers . . . against granting the petition of the Sect called Erika or Purists, who prayed for the abolition of piracy and slavery."[108] Europeans and Americans had long assailed north African Arabs for enslaved Christians captured in raids and at sea. In his published letter, Franklin related this fictional Arab's purported speech, which mirrored Jackson's in every respect except that the slaves were white instead of Black and the masters Muslim rather than Christian. The transposition may have fooled some readers; it certainly caught their eye and focused them on the core issue.

"Have these Erika considered the consequences of granting their petitions?" Franklin quoted Ibrahim as asking. "If we cease our cruises against the Christians, how shall we be furnished with the commodities their countries produce, and which are so necessary for us? If we forbear to make slaves of their people, who, in this hot climate, are to cultivate our lands." The speech went on, "And for what? to gratify the whim of a whimsical sect! who would have us not only forbear making more slaves, but even to manumit those we have."[109]

In words mocking Jackson, the speaker asked, "If we set our slaves free, what is to be done with them? . . . Our people will not pollute themselves by intermarrying with them: must we maintain them as beggars?" Then came Jackson's defense of slavery

as a positive good. "And what is so pitiable in their present condition? Were they not slaves in their own countries?" the speaker
mused. "Is their condition then made worse by their falling into
our hands? No, they have only exchanged one slavery for another:
and I may say a better: for here they are brought into a land where
the sun of Islamism gives forth its light, and shines in full splendor,
and they have an opportunity of making themselves acquainted
with the true doctrine, and thereby saving their immortal souls."
Modern satire often relies on cultural relativism, and here Franklin displayed his modernity. "While serving us, we take care to
provide them with every thing," he had Ibrahim say about Arabs
and their European slaves. "The labourers in their own countries,
are, as I am informed, worse fed, lodged and clothed." The speech
concluded by affirming that the Qur'an condones slavery in a passage that Franklin drew from a biblical verse, Ephesians 6:5, cited
by Jackson: "Slaves serve your masters with cheerfulness and
fidelity."[110] No reader could mistake Franklin's meaning.

America's first great humorist and essayist, as well as its leading
diplomat, scientist, inventor, and popular philosopher, Franklin
would pass away within a month of publishing this satire. "If to
be venerated for benevolence—if to be admired for talent—if to
be esteemed for patriotism—if to be beloved for philanthropy can
gratify the human mind, you must have the pleasing consolation
to know that you have not lived in vain," Washington had written to Franklin the previous fall, upon learning of his diminished
physical condition.[111]

For twenty-first-century Americans, Franklin's final crusade
served to confirm his benevolent, philanthropic, and forward-
looking nature. The nation's oldest founding father, Franklin also
seems the most modern—the one who would feel at home in
America today—while Washington's attachment to slavery feeds

his image as a marble icon from the past—a great but dated man. Franklin captured the eternally optimistic, fundamentally progressive, and essentially American aspects of his character when he wrote in one of his final letters to an old friend, "It is pleasant to see the World growing better and happier, tho' one is about to quit it."[112] His optimism extended to his own future as well for, as he wrote in a two-sentence note to his worried sister, "With respect to the Happiness hereafter which you mention I have no Doubts about it, confiding as I do in the Goodness of that Being who thro' so long a Life has conducted me with so many Instances of it."[113] Physically, as he assured his daughter shortly before his death, he was ready to go, but mentally—never.

Franklin died at home from a ruptured abscess in his lungs on April 17, 1790, with his oldest grandchildren, Temple and Benny, holding his hands.

Upon hearing of Franklin's death, Madison moved that members of the House of Representatives wear black badges for a month. The motion passed unanimously. In the Senate, however, where Vice President John Adams presided, a similar motion was withdrawn and, when Jefferson proposed that the executive branch follow the House, Washington demurred. "He said he would not know where to draw the line, if once he began that ceremony," Jefferson reported. "I told him the world had drawn so broad a line between himself & Dr. Franklin on the one side, and the residue of mankind on the other, that we might wear mourning for them, and the question still remain new & undecided as to all others."[114] A more logical reason for Washington's response—circulated then and repeated ever after—attributed the coolness to pique following Franklin's antislavery petition, which Washington depicted as "an ill-judged piece of business."[115] Yet once the seat of government moved from New York to Philadelphia later in

1790, Washington attended a eulogy for Franklin at the American Philosophical Society.

When news of Franklin's death reached France, the National Assembly decreed three days of mourning and a flood of tributes ensued. "The name of Benjamin Franklin will be immortal in the records of freedom and philosophy," the assembly's president declared in a letter to Washington.[116] No foreigner had ever received such honors in France. Acknowledging the decree by the assembly, Washington replied to its president, "So peculiar and so signal an expression of the esteem of that respectable body for a citizen of the United States, whose eminent and patriotic services are indelibly engraved on the minds of his countryman cannot fail to be appreciated."[117] In Philadelphia, nearly twenty thousand mourners turned out for Franklin's funeral procession—more than had ever gathered in the city for such an event. Franklin was the rare prophet honored both at home and abroad.

THE WALKING STICK

IN A CODICIL TO HIS WILL drafted soon after Washington became president, Franklin wrote, "My fine Crab tree walking stick with a gold head curiously wrought in the form of the cap of Liberty I give to my friend & the friend of Mankind General Washington—If it were a sceptre, he has merited it, & would become it."[1] Marianne Cammasse, Countess of Frobach, had given Franklin the "thorn stick," as she called it, or *le baton de Pommier savage*," as Franklin referred to it after receipt, in celebration of America's independence.[2] That a humble crab-apple staff could serve as an American leader's scepter carried a meaning that Franklin surely appreciated. Presidents would rule by popular assent rather than military might or divine right. Topping it with a liberty cap rather than a crown reinforced the symbolism. American leaders are called to secure the people's liberty, not their own aggrandizement. In a republican twist, Franklin suggested that, by wielding it, Washington gave authority to the scepter rather than its giving authority to him. Hailing Franklin as "great and invaluable," Washington commented on receiving the bequest, "As a token of remembrance and a mark of friendship, I receive this legacy with pleasing sensations and a grateful heart."[3]

It seems altogether fitting for the Sage of Philadelphia to invest Washington with a wild-apple-tree scepter in a manner akin to the ancient Hebrew prophet Samuel's anointing Saul as Israel's first king. Familiar with scripture and conversant in biblical history, Franklin and Washington surely recognized the similarity. That each invoked their friendship in extending and receiving the scepter makes the ritual particularly American and distinctly republican. As a later generation of Americans might say, Franklin literally and figuratively passed *"le baton"* to Washington, and Washington gratefully accepted it.

Assuming leadership of the world's first extended republic, Washington adopted the title "Mr. President" rather than the more regal "Your Majesty" or "His Most Benign Highness" proposed by Vice President John Adams.[4] To his friends, however, it remained General Washington and he continued signing his letters to Franklin as Go: Washington.

AS THE FIRST PRESIDENT, Washington brought the full force of his personality and popularity to the controversial task of forming a federal union from diverse states. A conciliator by nature who favored leading by consensus, he chose prominent leaders from a spectrum of those who had supported ratification to head the executive departments and to form a cabinet of advisors. For the critical domestic policy post of treasury secretary, he turned to Alexander Hamilton, who would push the Constitution's limits in creating a nation. For the State Department—the former office of foreign affairs—Washington picked the states-rights-minded Thomas Jefferson, who (much like Franklin) questioned the extent of federal and presidential power under the Constitution. Another Virginian, the off-and-on supporter of ratification Edmund Randolph, became attorney general. As secretary of war, Washington tapped his second in command from the Revolution, the ardent federalist Henry Knox. In dealing with Congress, Washington at first relied most heavily on James Madison in the House and Robert Morris in the Senate. To launch the judicial branch, Washington nominated New York federalist John Jay.

The results transformed America. Working almost as a team under Washington, the executive, legislative, and judicial branches began the process of forging a unified republic. The principal policies for doing so came from Hamilton who, much to the dismay of Jefferson and the despair of antifederalists, emerged as Washington's most influential advisor. However, Jefferson made his contributions to the emerging order, such as by devising a broad regime of federally protected intellectual property rights. With Knox, he also backed Washington's efforts to open the Ohio Country for settlement, leading to prolonged warfare with the Western Confederacy of Native tribes until its capitulation in 1795. Madison took the lead in pushing a Bill of Rights through Congress. Although the final amendments were less protective of states' rights than antifederalists wanted, they passed Congress in 1789 and Washington personally forwarded them to the states for ratification.

Hamilton's nationalizing policies were founded on funding the full debt run up by Congress and the states during the Revolutionary War. He viewed this as a means to align the interests of wealthy Americans with the central government, displace the states as independent economic actors, and enhance the country's credit. To pay for it, Hamilton pushed a tariff on imported goods, which would have the side effect of sheltering American industry, and an excise tax on such domestic items as whiskey, which he saw as a means to exert authority over frontier distillers.

As a capstone for his economic program, Hamilton wanted a quasi-independent central bank for the United States, co-owned by private investors, which would in effect regulate fiscal policy and provide a stable national currency. Here, Jefferson drew the line, claiming the Constitution did not authorize Congress to charter a bank. Invited by Washington to reply, Hamilton countered that the Necessary and Proper Clause, already reviled by antifederalists, authorized Congress to do virtually anything that advanced its express powers to lay taxes and regulate interstate commerce. Washington sided with Hamilton, and the bank was chartered in 1791. Two distinct factions had emerged, with the leaders of each—Hamilton and Jefferson—in Washington's cabinet.

Endorsed by both camps and encouraged to run as the only one able to bridge the growing partisan divide, Washington was unanimously elected to a second four-year term in 1792. Under the façade of unity, however, two distinct political parties were forming. It showed in a growing tendency for federal and state-level office seekers to align themselves with one of the two factions and then adhere to the party line if elected. Jefferson's partisans began calling themselves Republicans while Hamilton's followers formalized their designation as Federalists. The appearance of local newspapers closely identified with each of the opposing parties hardened divisions everywhere, with Franklin's grandson Benny publishing the Republicans' flagship paper, the *Aurora*, out of his grandfather's Market Street printing office.

Events pushed partisanship during Washington's second term. Frustrated by Hamilton's domination over the administration, Jefferson left the cabinet in 1793. A year later, Republicans denounced the government for suppressing resistance to the whiskey tax in western Pennsylvania with a thirteen-thousand-soldier army personally led by Washington. Then convulsions caused by the French Revolution and ensuing war between republican France and royalist Britain engulfed domestic politics. Washington's decision to proclaim neutrality without consulting Congress outraged Republicans, who viewed the United States as bound by Franklin's treaty to support its Revolutionary War ally. Here was presidential excess, they charged. When the Royal Navy nevertheless seized hundreds of American merchant ships bound for French ports in the West Indies, and impressed American sailors to boot, many Republicans favored a second war with Britain. Instead, in 1794, Washington sent Chief Justice Jay to resolve differences between the United States and its former colonial master. Bargaining from a weak position, Jay's treaty did little more than accept British limits on American trade with France in exchange for seemingly meaningless concessions. For the first time, Washington's popularity sagged. He was excoriated in the Republican press, and permanently embittered.

IN 1796, AT AGE SIXTY-FOUR, Washington announced that he would not accept a third term as president. He wished to retire, again, to Mount Vernon. His farewell address, printed in newspapers as a letter to the people, denounced partisanship, embraced economic nationalism, and discouraged permanent foreign alliances. Speaking to all Americans, Washington wrote, "The unity of Government which constitutes you one people, is also now dear to you. It is justly so; for it is a main pillar in the edifice of your real independence, the support of your tranquility at home, your peace abroad; of your safety; of your prosperity; and of that very Liberty which you so highly prize."[5] This address joined his circular letter of 1783 as one of Washington's two most significant public writings.

By this point, Washington's legacy as president was secure. His initiatives had restored American credit, opened the Ohio Country for settlement, fostered economic expansion, and established the presidency as a powerful office of overarching significance. Perhaps most important of all, he had kept the United States at peace with Europe during a period of widening transatlantic war. In the election of 1796, Federalist John Adams narrowly beat Thomas Jefferson in America's first contested election for president. Under the Constitution as then written, by coming in second, Jefferson became vice president, and the partisan split only deepened, especially after the worsening conflict in Europe led Adams and the Federalist-led Congress into waging a costly naval war with France and passing a series of laws to suppress domestic dissent, including the much maligned Alien and Sedition Acts that Republicans denounced as unnecessary, unwise, and unconstitutional. They would ride them to victory in the tumultuous election of 1800.

Back at his beloved Mount Vernon in 1797, Washington threw himself into farming and even became a whiskey distiller. No product ever netted him a larger return on his investment than this potent, rye-based intoxicant that he sold straight from the still. His distillery became the largest in the United States by 1799. On December 12 of that year a heavy snow started falling during Washington's daily ride around his plantation's five farms. He returned home wet and cold. By the thirteenth, a Friday, Washington's trusted aide Tobias Lear reported about his longtime employer, "He had taken cold (undoubtedly from being so much exposed the day before) and complained of having a sore throat."[6]

Washington's condition grew worse by the next morning. He struggled for breath and could scarcely speak. To counter the inflammation that strangled him, he asked for a bleeding by the overseer who generally treated Mount Vernon's slaves. When they arrived, Washington's doctors repeated the procedure three more times and administered two laxatives. Nothing helped. In all likelihood Washington had contracted epiglottitis, which no available medical treatment could cure. He accepted his fate.

After reviewing two draft wills that he had prepared in anticipation of his demise, Washington confirmed the one that would free his slaves upon his wife's death, had the other destroyed, and waited for his passing.

"Doctor, I die hard, but am not afraid to go," he said at dusk. The end came late that night. His final words were, "'Tis well."[7]

News of the unexpected death touched off an outpouring of grief unprecedented in the country's history. "Every paper we received from towns which have heard of Washington's death, are enveloped in mourning," one journalist reported near the year's end. "Every city, town, village and hamlet has exhibited spontaneous tokens of poignant sorrow."[8]

DELAYING UNTIL DEATH to reveal his intention of freeing his slaves, and then postponing their release still longer, drained the act of its potential political and social significance. Trusted aides (like Lafayette) and prominent abolitionists (including Virginia Quaker Robert Pleasants) had urged Washington to act earlier—ideally during the idealistic fervor of the American Revolution—when it might have made a difference. "Remember the cause for which thou wert call'd to the Command of the American Army, was the cause of Liberty and the Rights of Mankind," Pleasants wrote to Washington in 1784. "How strange then must it appear to impartial thinking men . . . that thou . . . should now withhold that enestimable blessing from any who are absolutely in thy power."[9] Other chances for Washington to take a public stand came at the Constitutional Convention or, along with Franklin, upon the founding of the federal republic.

Once dead, Washington could neither explain his motives nor present his final act as a model for other slaveholders. Even his wife, who owned most of the slaves at Mount Vernon as her dower property, did not follow his lead and kept even the mixed-race children allegedly sired by her father (her half sisters) and her son (her grandchildren) enslaved. Southern slaveholders easily dismissed Washington's deathbed act and northern abolitionists struggled to give it meaning.

In any event, by 1800, the moment (if there ever was one) to end slavery in America without civil war had passed. The new federal republic facilitated the opening of rich farmland west of coastal Georgia for settlement. New technologies for processing the short-staple strains suitable for cultivation there greatly expanded southern cotton production—making it the nation's leading export and fueling the demand for slaves. Constitutional protection for the slave trade through 1808 and the development of textile production in the northeast under Hamilton's protective tariffs created ideal conditions for integrating slavery into the national market economy. Nothing Washington might do with slaves on his tidewater Virginia wheat farms could speak to the economics of slavery in the cotton south. Others would suffer for these sins.

No one can know what might have happened had the two icons of the Revolution, Franklin and Washington, stood together against slavery at the nation's founding. Certainly some of their contemporaries thought it could have made a difference. As it happened, they split over the issue and with them the nation. Washington and the southern states retained their slaves. Franklin and the northern states rejected the institution.

Despite Washington's decision to free his slaves after his death, during the antebellum period, southern slaveholders counted him among the founders who owned slaves. Upon secession, the official seal of the Confederate States of America featured the mounted figure of a uniformed George Washington at its heart. To launch the slaveholding republic, Jefferson Davis took the oath of office as the president of the Confederacy on Washington's birthday, 1862. After his surrender in 1865, Confederate commander

Robert E. Lee (who was married to Martha Washington's great-granddaughter) became president of a Virginia college named for Washington that, upon Lee's death, school officials renamed for Washington and Lee. The new name carried meaning. Postbellum architects of the Lost Cause of the Confederacy often linked Washington to the heroes of the Confederacy, such as when Lee's son George Washington Custis Lee reviewed the 1907 grand parade of Confederate veterans. More than a century later, objecting to the removal of a statue of Robert E. Lee in Charlottesville, Virginia, President Donald Trump sharply asked, "I wonder, is it George Washington next week?"[10] The question is stark. Its answer involves balanced assessments of each figure's contributions and character that successive generations perform anew.

PLEASANTS CLOSED HIS 1784 LETTER to Washington with a warning: "Notwithstanding thou art now receiving the tribute of praise from a grateful people, the time is coming when all actions will be weighed in an equal ballance, and undergo an impartial examination; how inconsistant then will it appear to posterity, should it be recorded, that the Great General Washington . . . [would] keep a number of People in absolute Slavery, who were by nature equally entitled to freedom as himself."[11] This same examination applies to all the founders.

Despite their flaws, Franklin and Washington have held up better under examination than most leaders of any age. Theirs was the founding partnership that launched a nation. Over the years, the harshest critics of Franklin have focused on his promotion of stultifying, middle-class virtues. "Early to bed and early to rise, makes a man healthy wealthy and wise," he wrote in an archetypal *Poor Richard's Almanack* aphorism for 1735, and added a year later, "Poverty, Poetry, and new Titles of Honour, make Men ridiculous."[12] Generations of Americans took Franklin's maxims to heart, with some like publisher James Harper and banker Thomas Mellon crediting them as their way to wealth. Just as surely, generations of intellectuals from Edgar Allan Poe and Herman Melville to F. Scott Fitzgerald and D. H. Lawrence mocked Franklin as a pedestrian prophet of pragmatism. Yet Franklin was a man of many faces, who as an author hid behind masks ranging from his first, the witty widow Silence Dogood, to his last, the Arab slaver Sidi Mehemet Ibrahim. Judging Franklin solely in his Richard Saunders guise fails to do him justice.

Washington did not wear multiple masks but so carefully cultivated the firm face of republican virtue that he believed his countenance never betrayed his true feelings.[13] He viewed favorable public opinion as a formidable but fickle foundation for leadership in a republic, and endeavored to foster it. Washington's success led the less popular Adams, with a mixture of esteem and envy, to view the general as an actor playing a role: a "Character of Convention," Adams called him, designedly made "popular and fashionable with all parties and in all places and with all persons as a center of union."[14] Indeed, after observing him up close as president, Adams described Washington as the finest political actor he ever witnessed in action.[15] This aspect of Washington's personality can make it as difficult to see behind his public image as it is to look beyond Franklin's multiple guises and middle-class manners. Then as now, the Pennsylvania printer and the Virginia planter appeared too dissimilar to establish and maintain a lasting friendship, especially since the former posed as a man of the people while the latter preened as one above them.

Yet focusing on their distinct public images obscures their fundamental similarities. Hardworking and entrepreneurial, Franklin and Washington had successful business careers outside government and never viewed themselves primarily as politicians. Both prospered as colonists and supported royal rule until realizing that Britain would not extend basic English rights to Americans. Jealous of their liberties, they turned against the crown and never looked back. Each nurtured deep, lifelong relationships with both men and women. Natural leaders, people trusted them and they trusted others. Both men listened more than they talked, compromised on means to secure ends, relied on others, sacrificed for the common good, and never wavered on principle. And both were reformers—Franklin compulsively so. They saw problems and tried to fix them. Franklin's fixes ranged from mechanical to moral—lightning rods, stoves, and bifocals to constitutions, ethical codes, and popular philosophy. Washington's included constitutions, of course, but also military and agricultural reforms.

Shaped by the Enlightenment, Franklin and Washington shared a republican ideology and progressivist faith that relied on human reason and divine providence rather than traditional ways and established dogmas. They sought truth and accepted facts. Life could get better, they believed. Theirs did.

As the old order collapsed around them, they crafted a better one to replace it—one that has lasted for more than two centuries. They did not see it as perfect and never thought it would last forever. If the people allowed it, Franklin warned, even the Constitution, for all its virtues, would lead to tyranny, with the presidency serving as "the fetus of a King."[16] The example of Franklin and Washington, however, shows what individuals can do in times of faction, fracture, and failure to address problems and improve the state of affairs. "We will not be driven by fear," the legendary broadcast journalist Edward R. Murrow would later say about Americans, "if we dig deep in our history and our doctrine and remember that we are not descended from fearful men."[17] Murrow surely had the likes of Franklin and Washington in mind. And so, at the onslaught of World War II—the war that made Murrow famous—in his Four Freedoms speech, a resolute Franklin D. Roosevelt quoted Franklin, "Those who would give up essential liberty, to purchase a little temporary safety, deserve neither liberty nor safety."[18]

Franklin was more than a Philadelphia printer. Washington much more than a tidewater planter. They were larger-than-life American originals whose partnership in revolutionary times laid the foundation for the world's first continental republic, which has lasted for nearly 250 years. Each recognized the other's goodness and greatness, and they viewed one another as partners in the fight for liberty. Others saw this too. Despite their critics, Franklin was elected to his state's highest office unanimously and Washington was elected to his nation's highest office unanimously. Central to their republican conception of service, both men willingly relinquished their public stations to return to their private positions. Indeed, both preferred private life to public power. For each, so long as it had a liberty cap instead of a crown, a crab tree walking stick served as a fitting scepter.

NOTES

List of Abbreviations for Frequently Cited Works

AFC: L. H. Butterfield et al., eds., *Adams Family Correspondence,* 13 vols. (Cambridge: Harvard University Press, 1963–).

Annals of Congress: *The Debates and Proceedings in the Congress of the United States,* 42 vols. (Washington: Gales and Seaton, 1834).

DAJA: L. H. Butterfield et al., eds., *Diary and Autobiography of John Adams,* 4 vols. (Cambridge: Harvard University Press, 1961).

DGW: Donald Jackson and Dorothy Twohig, eds., *The Diaries of George Washington,* 6 vols. (Charlottesville: University Press of Virginia, 1976–1979).

DHFFC: Linda Grant DePauw et al., eds., *The Documentary History of the First Federal Congress of the United States of America,* 22 vols. (Baltimore: Johns Hopkins University Press, 1972–).

DHRC: Merrill Jensen et al., eds., *The Documentary History of the Ratification of the Constitution,* 27 vols. (Madison: State Historical Society of Wisconsin, 1976–).

Farrand: Max Farrand, ed., *The Record of the Federal Convention of 1787,* 4 vols., rev. ed. (New Haven: Yale University Press, 1937).

JCC: Library of Congress, *Journals of the Continental Congress, 1774–1789,* 34 vols. (Washington, D.C.: Government Printing Office, 1904–1937).

PAH: Harold C. Syrett and Jacob Cooke, eds, *The Papers of Alexander Hamilton,* 27 vols. (New York: Columbia University Press, 1961–1987).

PBF: Leonard W. Labaree et al., *The Papers of Benjamin Franklin,* 43 vols. (New Haven: Yale University Press, 1959–). The first forty-three volumes in this series, running through March 15, 1785, have been published as of 2019. The unpublished volumes are available unpaginated online at franklinpapers.org and are cited here by date and projected volume number with page numbers left blank.

PGM: Robert A. Rutland, ed., *The Papers of George Mason, 1725–1792,* 3 vols. (Chapel Hill: University of North Carolina Press, 1970).

PGW-CfS: W. W. Abbot et al., eds., *The Papers of George Washington: Confederation Series*, 6 vols. (Charlottesville: University Press of Virginia, 1992–1997).

PGW-CS: W. W. Abbot et al., eds., *The Papers of George Washington: Colonial Series*, 10 vols. (Charlottesville: University Press of Virginia, 1983–1995).

PGW-PS: W. W. Abbot et al., eds., *The Papers of George Washington: Presidential Series*, 19 vols. (Charlottesville: University Press of Virginia, 1987–).

PGW-RS: Dorothy Twohig et al., eds., *The Papers of George Washington: Retirement Series*, 4 vols. (Charlottesville: University Press of Virginia, 1998–1999).

PGW-RWS: W. W. Abbot et al., eds., *The Papers of George Washington: Revolutionary War Series*, 26 vols. (Charlottesville: University Press of Virginia, 1985–).

PJA: Robert J. Taylor et al., eds., *Papers of John Adams*, 19 vols. (Cambridge: Harvard University Press, 2003–). The first nineteen volumes in this series, running through May 1789, have been published as of 2019. The unpublished volumes are available unpaginated online at the Adams Papers section of founders.archives.gov and are cited here by date with volume and page numbers left blank.

PJM: Robert A. Rutland et al., eds., *The Papers of James Madison*, 17 vols. (Chicago and Charlottesville: University of Chicago Press and University Press of Virginia, 1962–).

PTJ: Julian P. Boyd et al., eds., *The Papers of Thomas Jefferson*, 43 vols. (Princeton, N.J.: Princeton University Press, 1950–).

PTJ-RS: Jefferson Lobney et al., eds., *The Papers of Thomas Jefferson, Retirement Series*, 15 vols. (Princeton, N.J.: Princeton University Press, 2005–). The first fifteen volumes in this series, running through May 31, 1820, have been published as of 2019. The unpublished volumes are available unpaginated online at the Jefferson Papers section of founders.archives .gov and are cited here by date with volume and page numbers left blank.

WGW: John Clement Fitzpatrick et al., eds., *The Writings of George Washington from the Original Manuscript Sources, 1745–1799*, 39 vols. (Washington, D.C.: Government Printing Office, 1931–1944).

Preface: "My Dear Friend"

1. Benjamin Franklin to Thomas Jefferson, April 19, 1787, PBF, 46:---.
2. E.g., George Washington to Benjamin Harrison, September 24, 1787,

PGW-CfS, 5:339. In this letter, Washington wrote that "the political concerns of this Country are, in a manner, suspended by a thread" and that if the Convention had not agreed on the Constitution, "anarchy would soon have ensued." On the same day, Washington sent virtually identical letters to former Virginia governors Patrick Henry and Thomas Nelson.

3. Benjamin Franklin to Jane Mecom, September 21, 1786, PBF, 45:---.
4. Thomas Jefferson to Benjamin Rush, October 4, 1803, PTJ, 41:471.

Chapter One: Great Expectations

1. Gilbert Stuart and George Washington, quoted in James MacGregor Burns and Susan Dunn, *George Washington* (New York: Times/Holt, 2004), 58.
2. William Pierce, "Character Sketches of the Delegates to the Federal Convention," in Farrand, 3:91.
3. George Washington, quoted in Paul K. Longmore, *The Invention of George Washington* (Berkeley: University of California Press, 1988), 181.
4. Gilbert Stuart, quoted in Carrie Rebora Barratt and Ellen G. Miles, *Gilbert Stuart* (New York: Metropolitan Museum of Art, 2004), 137.
5. H. W. Brands, *The First American: The Life and Times of Benjamin Franklin* (New York: Doubleday, 2000), 8.
6. Joseph J. Ellis, *Founding Brothers: The Revolutionary Generation* (New York: Random House, 2000), 53, 163.
7. Gary Wills, *James Madison* (New York: Times, 2002), 164; Gordon S. Wood, *The Americanization of Benjamin Franklin* (New York: Penguin Press, 2004), 54.
8. Wood, *Americanization of Franklin*, 246.
9. John Adams to Benjamin Rush, April 4, 1790, Founders Online, National Archives, Washington, D.C. Eighteenth-century spelling and capitalization in quotations left as per original here and throughout.
10. Silence Dogood [Benjamin Franklin] to the Author of the *New-England Courant*, *New-England Courant*, April 16, 1722, in PBF, 1:12–13.
11. Benjamin Franklin, *The Autobiography* (New Haven: Yale University Press, 1964), 119.
12. Ibid., 125–26.
13. Ibid., 88.
14. [Benjamin Franklin], "Dialogue Between Two Presbyterians," *Pennsylvania Gazette*, April 10, 1735, in PBF, 2:30.

15. [Benjamin Franklin], *Poor Richard, 1739. An Almanack For the Year of Christ 1739,* in ibid., 2:224.

16. Franklin, *Autobiography,* 115.

17. Benjamin Franklin, "Advice to a Young Tradesman, Written by an Old One," July 21, 1748, in PBF, 3:308.

18. Benjamin Franklin, "Old Mistresses Apologue," June 25, 1745, in ibid., 3:27. In this satirical writing, beyond his seemingly sincere comments on marriage, Franklin argued the practical advantages of taking an old mistress to a young one without any reference to physical attraction.

19. On Franklin's wealth, see Wood, *Americanization of Franklin,* 54.

20. Benjamin Franklin to Abiah Franklin, April 12, 1750, PBF, 3:475.

21. Franklin, *Autobiography,* 197.

22. Benjamin Franklin, "Opinions and Conjectures Concerning the Properties and Effects of the Electrical Matter, Arising from Experiments and Observations Made in Philadelphia, 1749," July 29, 1750, in PBF, 4:13.

23. Carl van Doran, *Benjamin Franklin* (New York: Viking, 1961), 171.

24. Franklin, *Autobiography,* 153.

25. For the new or modern Prometheus quote, see Walter Isaacson, *Benjamin Franklin: An American Life* (New York: Simon & Schuster, 2003), 145; Claude-Anne Lopez and Eugenia W. Herbert, *The Private Franklin: The Man and His Family* (New York: Norton, 1975), 47; van Doran, *Franklin,* 171.

26. For an analysis of Franklin's college admissions prospects, see Isaacson, *Franklin,* 18–20.

27. George Washington, "Remarks," 1787–1788, in PGW-CfS, 5:516.

28. George Washington to Marquis de Lafayette, December 8, 1784, ibid., 2:175.

29. Ibid.

30. George Washington, "A Journal of My Journey over the Mountains Begun Fryday the 11th of March 1747/8," in DGW, 1:10.

31. Ibid., 1:12.

32. Ibid., 1:13.

33. On Washington as a Freemason, see Ron Chernow, *Washington: A Life* (New York: Penguin Press, 2010), 28.

34. "The Second Charter of Virginia," May 23, 1609, in *The Federal and State Constitutions, Colonial Charters, and Other Organic Laws of the States, Territories, and Colonies Now or Heretofore Forming the United States of America,* ed. Francis Newton Thorpe (Washington, D.C.: Government Printing Office, 1909), 7:3795.

35. Earl of Holderness to Robert Dinwiddie, August 28, 1753, DGW, 1:126.

36. Washington, "Remarks," in PGW-CS, 5:116 (Washington's remarks accepting these words written by his aide David Humphreys, about the reaction of some to his choice for this mission).

37. George Washington, "The Journal of Major George Washington," in DGW, 1:137.

38. Robert Dinwiddie to French Commandant, October 30, 1753, ibid., 1:127; Legardeur de St. Pierre to Robert Dinwiddie, December 15, 1753, ibid., 1:151.

39. Washington, "The Journal," December 14, 1753, in ibid., 1:150.

40. Washington, "The Journal," December 15, 1753, in ibid., 1:151.

41. Christopher Gist, "Diary," December 21 and 22, 1753, in ibid., 1:154n60.

42. Washington, "The Journal," December 23, 1753, in ibid., 1:154.

43. Ibid.

44. Washington, "The Journal," December 15, 1753, in ibid., 1:152.

45. Washington, "The Journal," December 23, 1753, in ibid., 1:155.

46. Ibid.

47. Gist, "Diary," December 27, 1753, in ibid., 1:157n65.

48. Washington, "The Journal," December 23, 1753, in ibid., 1:155.

49. Gist, "Diary," December 29, 1753, in ibid., 1:158n65.

50. Joseph J. Ellis, *His Excellency: George Washington* (New York: Random House, 2004), 7.

51. James Thomas Flexner, *George Washington: The Forge of Experience* (Boston: Little, Brown, 1965), 80.

52. Benjamin Franklin, "Observations Concerning the Increase of Mankind, Peopling of Countries, &c., 1751," in PBF, 4:228, 233.

53. E.g., "Williamsburg, February 8," *Pennsylvania Gazette,* March 12, 1754, 1–2; "The Commandant's Answer," *Pennsylvania Gazette,* March 26, 1754, 1 (printed in French and English); "Charles-Town (in South Carolina), March 26," *Pennsylvania Gazette,* May 2, 1754, 1 (quoted passage).

Chapter Two: Lessons from the Frontier

1. George Washington, "Journal of the Expedition to the Ohio," April 20, 1754, in DGW, 1:177.

2. George Washington to James Hamilton, April [24], 1754, PGW-CS, 1:83; George Washington to Horatio Sharpe, April 24, 1754, ibid., 1:86.

3. George Washington to Robert Dinwiddie, March 9, 1754, ibid., 1:73–74.

4. George Washington to Robert Dinwiddie, May 27, 1754, ibid., 1:105.

5. George Washington to John Augustine Washington, May 31, 1754, ibid., 1:118.

6. George Washington, "The Journal of George Washington," May 26, 1754, in ibid., 1:197–98.

7. Tanacharison, as reported in the journal of Conrad Weiser, in Stephen Brumwell, *George Washington: Gentleman Warrior* (New York: Quercus, 2012), 57.

8. G. Washington to J. A. Washington, May 31, 1754, 1:118.

9. "Articles of Capitulation," July 3, 1754, in PGW-CS, 1:166.

10. Ibid., 1:165.

11. Marquis de Duquesne to Seigneur de Contrecoeur, September 8, 1754, DGW, 1:172.

12. [Benjamin Franklin], "Philadelphia, May 9," *Pennsylvania Gazette,* May 9, 1754, 2.

13. Benjamin Franklin to James Parker, March 20, 1751, PBF, 4:119.

14. Walter Isaacson, *Benjamin Franklin: An American Life* (New York: Simon & Schuster, 2003), 160–61.

15. Benjamin Franklin to William Shirley, December 4, 1754, PBF, 5:444.

16. Benjamin Franklin, *The Autobiography* (New Haven: Yale University Press, 1964), 210.

17. Ibid., 211.

18. Edward Braddock to Thomas Robinson, June 5, 1755, PBF, 6:14.

19. Benjamin Franklin, "Advertisement," April 26, 1755, in ibid., 6:20–22.

20. William Shirley Jr. to Robert Hunter Morris, May 14, 1755, ibid., 6:22n3.

21. Franklin, *Autobiography,* 223–24.

22. George Washington to William Fitzhugh, November 15, 1754, PGW-CS, 1:226.

23. George Washington to John Augustine Washington, June 28–July 2, 1755, ibid., 1:319, 324.

24. Philip Hughes to correspondent, July 23, 1955, in *Public Advertiser* (London), October 31, 1755, reprinted in N. Darnell Davis, "British Newspaper Accounts of Braddock's Defeat," *Pennsylvania Magazine of History and Biography* 23 (1899): 324.

25. George Washington to John Augustine Washington, July 18, 1755, PGW-CS, 1:343.

26. Adam Stephen to John Hunter, July 18, 1755, in Fred Anderson, *Crucible of War: The Seven Years' War and the Fate of Empire in British North America, 1754–1766* (New York: Knopf, 2001), 103.

27. George Washington to Robert Dinwiddie, July 18, 1755, PGW-CS, 1:339.

28. Ibid., 1:339–40.

29. George Washington, "Biographical Memorandum, c. 1789," in ibid., 1:332–33.

30. G. Washington to Dinwiddie, July 18, 1755, 1:340.

31. Benjamin Franklin to Jared Eliot, August 31, 1755, PBF, 6:172.

32. Jean-Daniel Dumas to the Minister, July 24, 1755, in Paul A. W. Wallace, *Conrad Weiser: Friend of Colonist and Mohawk* (Philadelphia: University of Pennsylvania Press, 1945), 385.

33. "The Governor's Speech," July 24, 1755, in *Minutes of the Provincial Council of Pennsylvania* (Harrisburg: Fenn, 1851), 6:487.

34. [Benjamin Franklin], Reply to the Governor, August 8, 1755, in PBF, 6:138.

35. [Benjamin Franklin], Reply to the Governor, August 19, 1755, in ibid., 6:162.

36. [Benjamin Franklin], Reply to the Governor, November 11, 1755, in ibid., 6:242.

37. George Glewell et al. to Robert Hunter Morris, October 20, 1755, ibid., 6:648.

38. Peter Spycker to Conrad Weiser, November 16, 1755, in Wallace, *Conrad Weiser,* 410.

39. Robert Hunter Morris to Colonial Governors, [November 1755], ibid., 413.

40. Benjamin Franklin to Peter Collinson, August 27, 1755, PBF, 6:169.

41. See, e.g., Thomas Lloyd to [John Hughes], January 30, 1756, ibid., 6:381.

42. Benjamin Franklin to Capt. Vanetta, January 12, 1756, PBF, 6:354.

43. For commentary on Franklin's military success, see J. A. Leo Lemay, *The Life of Benjamin Franklin* (Philadelphia: University of Pennsylvania Press, 2009), 3:514.

44. Christopher Gist to George Washington, October 15, 1755, PGW-CS, 2:115.

45. Robert Dinwiddie to George Washington, Commission, August 14, 1755, ibid., 2:4.

46. George Washington, "Orders," October 6, 1755, in ibid., 2:76.

47. George Washington to Adam Stephen, February 1, 1756, ibid., 2:310.

48. Benjamin Franklin to Peter Collinson, November 5, 1756, PBF, 7:13–15.

49. George Washington to Robert Dinwiddie, October 11, 1755, PGW-CS, 2:102.

50. Benjamin Franklin to George Washington, August 19, 1756, PBF, 6:488.

51. Adam Stephen to George Washington, October 4, 1755, PGW-CS, 2:72.

52. George Washington to Robert Dinwiddie, April 19, 1756, PGW-CS, 3:20.

53. George Washington, "Remarks," 1787–1788, in PGW-CfS, 5:524–25.

54. Benjamin Franklin to Thomas Pownall, August 19, 1756, PBF, 6:487.

55. Benjamin Franklin, "Observations Concerning the Increase of Mankind, Peopling of Countries, &c., 1751," in PBF, 4:234.

56. George Washington to Robert Dinwiddie, March 10, 1757, PGW-CS, 4:113. For a second such petition related to this conference, see George Washington to Lord Loudoun, March 23, 1757, ibid., 4:120–21.

57. Thomas Penn to Richard Peters, May 14, 1757, PBF, 7:111n9.

58. For a similar conclusion, see Ron Chernow, *George Washington: A Life* (New York: Penguin, 2010), 92–93.

59. Franklin, *Autobiography,* 226.

Chapter Three: From Subjects to Citizens

1. The term "Sons of Liberty" is attributed to Isaac Barré, quoted in Jared Ingersoll to Thomas Finch, February 11, 1765, in "A Selection from the Correspondence and Miscellaneous Papers of Jared Ingersoll," in *Papers of the New Haven Colony Historical Society* (New Haven: Tuttle, Morehouse, and Tayler, 1918), ed. Franklin B. Dexter, 9:312.

2. George Washington to James Gildart, April 26, 1763, PGW-CS, 7:201.

3. "Council and Burgesses of Virginia to King George III," December 18, 1754, *Journals of the House of Burgesses, 1761–65* (Richmond: Virginia State Library, 1907), 302.

4. "Memorial of the Council and Burgesses of Virginia," December 18, 1754, in ibid., 302.

5. Ingersoll to Finch, February 11, 1765, 9:312.

6. Benjamin Franklin to Joseph Galloway, October 11, 1766, PBF, 13:448.

7. Barré, quoted in Ingersoll to Finch, February 11, 1765, 9:311; "Diary of Nathanial Ryder," in *Proceedings of the Debates of the British Parliament Respecting North America, 1754–1783,* eds. R. C. Simmons and Peter D. G. Thomas (Millwood, N.Y.: Kraus, 1983), 2:13 (punctuation altered).

8. This version of the legend is from the printed caption of "Patrick Henry Delivers His Celebrated Speech," a popular nineteenth-century engraving by Alfred Jones, after a painting by P. F. Rothermel, c. 1852, copy in the Library of Virginia, Richmond, Virginia.

9. The text for these resolutions is from the July 4, 1765, issue of the *Mary-*

land Gazette and, along with a discussion of variations in the text and content of the Virginia resolves, appears in the revised edition of the classic account of this episode, Edmund J. Morgan and Helen M. Morgan, *The Stamp Act Crisis: Prologue to Revolution,* rev. ed. (New York: Collier, 1963), 120–32 (text from *Gazette* on 128–29). For the official text of the first four resolutions as they appear in the assembly journal, see "Thursday, the 30th of May, 5 Geo. III, 1765," in *Journals of the House of Burgesses of Virginia, 1761–1765,* ed. John Pendleton Kennedy (Richmond: State Library, 1907), 360.

10. Deborah Franklin to Benjamin Franklin, September 22, 1765, PBF, 12:271.

11. Benjamin Franklin to Deborah Franklin, November 9, 1865, PBF, 12:361.

12. George Washington to Robert Cary & Co., September 20, 1765, PGW-CS, 7:401–2.

13. Benjamin Franklin, "Examination Before the Committee of the Whole of the House of Commons," February 13, 1766, in PBF, 13:135.

14. Ibid., 13:142.

15. Ibid., 13:147.

16. Ibid., 13:144.

17. Ibid., 13:156.

18. Ibid., 13:141.

19. George Washington to Robert Cary & Co., July 21, 1766, PGW-CS, 7:457.

20. Franklin, "Examination," 13:150.

21. Writing as Poor Richard in his *Almanack,* Franklin often made comments to the effect that, to have a friend, be a friend; not to throw stones at your neighbor's if your own windows are glass; and people who sow thorns should not go barefoot.

22. "Friday, the 21st of November, 7 Geo. III, 1766," in *Journals of the House of Burgesses of Virginia, 1766–1769,* ed. John Pendleton Kennedy (Richmond: State Library, 1906), 34.

23. G. Washington to Cary, September 20, 1765, 7:401.

24. Franklin, "Examination," 13:143.

25. George Washington to George Mason, April 5, 1769, PGW-CS, 8:179. The passage quoted was written in the first person to apply to Virginia planters generally, but clearly applies to Washington himself as well.

26. Ibid., 8:178.

27. George Mason to George Washington, April 5, 1769, ibid., 8:182.

28. "Virginia Non-Importation Resolutions," May 17, 1769, in PTJ, 1:29. In contrast, the Philadelphia nonimportation agreement was addressed to

"The Merchants and Traders of the City of Philadelphia": see Stanislaus Murray Hamilton, ed., *Letters to Washington and Accompanying Papers* (Boston: Houghton Mifflin, 1901), 3:351.

29. George Washington to Robert Cary & Co., July 25, 1769, PGW-CS, 8:229.

30. "Wednesday, the 17th of May, 9 Geo. 1769," in *Journals of the House, 1766–1769*, 218.

31. "Thursday, May 18," in ibid., xliii. They also toasted the king, queen, and royal family, as well as Lord Botetourt.

32. Benjamin Franklin to Deborah Franklin, April 6, 1766, PBF, 13:233.

33. Benjamin Franklin to Jane Mecom, March 1, 1766, ibid., 13:188.

34. [John Dickinson], *Letters from a Farmer in Pennsylvania to the Inhabitants of the British Colonies* (London: Almon, 1768), 7, 17.

35. Benjamin Franklin to William Franklin, March 13, 1768, PBF, 15:75–76.

36. Benjamin Franklin to Samuel Cooper, June 8, 1770, ibid., 17:162–63.

37. Benjamin Franklin to William Franklin, October 6, 1773, ibid., 20:437.

38. Benjamin Franklin to Charles Thomson, March 18, 1770, ibid., 17:113.

39. Benjamin Franklin to William Straham, November 29, 1769, ibid., 16:244–45.

40. George Washington to Jonathan Boucher, July 30, 1770, PGW-CS, 8:361. For the modifications, compare "Virginia Resolutions," May 17, 1769, in PTJ, 1:29 with "Virginia Non-Importation Resolutions," June 22, 1770, in ibid., 1:44.

41. George Washington to Jonathan Boucher, July 30, 1770, PGW-CS, 8–361 (first quotation); George Washington to Robert Cary & Co., August 20, 1770, ibid., 8:371 (second quotation).

42. George III to Lord North, February 4, 1774, in *The Correspondence of King George the Third with Lord North from 1768 to 1783,* ed. W. Bodham Dunne (London: Murray, 1867), 1:164 (source of both quotations).

43. Thomas Hutchinson to ---, January 20, 1769, PBF, 20:550.

44. Thomas Hutchinson to ---, October 20, 1769, ibid., 20:551.

45. "Tuesday, the 24th of May, 14 Geo. III, 1774," in *Journals of the House of Burgesses of Virginia, 1773–1776,* ed. John Pendleton Kennedy (Richmond: State Library, 1905), 124.

46. "Association of Members of the Late House of Burgesses," May 27, 1774, in PTJ, 1:107–8.

47. "At the Committee of Correspondence," May 28, 1774, in Kennedy, ed., *Journals of the House, 1773–1776,* 138.

48. "Fairfax County Resolves," July 17, 1774, in PGW-CS, 10:121–22.

49. George Washington to Bryan Fairfax, July 20, 1774, ibid., 10:129–30.

50. George Washington to George William Fairfax, June 10–15, 1774, ibid., 10:96–97.

51. "Resolution at the Association of the Convention of Virginia of 1774," August 1–6, 1774, in PTJ, 1:139.

52. Ibid., 1:138; "Instructions by the Virginia Convention to Their Delegates in Congress, 1774," August 1774, in ibid., 1:143.

53. "Continental Association," October 20, 1774, in ibid., 1:149–54.

54. George Washington to Robert McKenzie, October 9, 1774, PGW-CS, 10:171–72.

55. Patrick Henry, "Give Me Liberty or Give Me Death," March 23, 1775, Avalon Project: Documents in Law, History, and Diplomacy, Yale Law School, https://avalon.law.yale.edu/18th_century/patrick.asp.

56. William Franklin to Benjamin Franklin, May 3, 1774, PBF, 21:207.

57. Benjamin Franklin to Jane Mecom, September 26, 1774, ibid., 21:317–18.

58. Benjamin Franklin, "Journal of Negotiations in London," March 22, 1775, in PBF, 21:583.

59. George Washington to George William Fairfax, May 31, 1775, PGW-CS, 10:368.

60. Washington made this distinction explicit when he wrote, "The Crisis is arrivd when we must assert our Rights, or Submit to every Imposition that can be heap'd upon us; till custom and use, will make us as tame, & abject Slaves, as the Blacks we Rule over with such arbitrary Sway." George Washington to Bryan Fairfax, August 24, 1774, ibid., 10:155.

61. John Adams, quoted in David Hackett Fischer, *Paul Revere's Ride* (New York: Oxford University Press, 1994), 279.

62. "Extract of a Letter from Philadelphia," May 8, 1775 (New York: John Anderson, 1775), Library of Congress, Washington, D.C., Broadsides, Leaflets, and Pamphlets from America and Europe, Portfolio 108, Folder 19.

Chapter Four: Taking Command

1. Arthur Lee, "Extracts from Journal," October 25, 1778, in Richard Henry Lee, *Life of Arthur Lee* (Boston: Wells & Lilly, 1829), 1:345.

2. Joseph J. Ellis, *His Excellency: George Washington* (New York: Random House, 2005), 84.

3. E.g., George Washington to Martha Washington, June 18, 1775, PGW-RWS, 1:3 ("You may beleive me my dear Patcy, when I assure you, in the most solemn manner, that, so far from seeking this appointment I have used every endeavour in my power to avoid it").

4.   John Adams to James Warren, July 24, 1775, PJA, 3:89. For Lee's response, see Charles Lee to John Adams, October 5, 1775, ibid., 3:186 ("Untill the bulk of Mankind is much alter'd I consider the reputation of being whimsical and eccentric rather as a panegyric than sarcasm and my love of Dogs passes with me as a still higher complement").

5.   Silas Deane to Elizabeth Deane, September 10–11, 1774, in *Letters of the Delegates to Congress, 1774–1789*, ed. Paul H. Smith (Washington, D.C.: Library of Congress, 1976), 1:61–62.

6.   JCC, June 14, 1775, 2:90.

7.   Samuel Ward to Henry Ward, October 5, 1775, in Smith, ed., *Letters of the Delegates*, 2:123.

8.   JCC, June 17, 1775, 2:96.

9.   "Franklin's Design of Paper Currency," in PBF, 22: illustration following p. 358.

10.   JCC, June 16, 1775, 2:92.

11.   G. Washington to M. Washington, June 18, 1775, 1:4.

12.   George Washington to John Augustine Washington, June 20, 1775, PGW-RWS, 1:19.

13.   Thomas Jefferson to Walter Jones, January 2, 1814, PTJ-RS, 7:101.

14.   Abigail Adams to John Adams, July 16, 1775, AFC, 1:246.

15.   Benjamin Rush to Thomas Rustin, October 29, 1775, in *Letters of Benjamin Rush,* ed. L. H. Butterfield (Princeton, N.J.: Princeton University Press, 1951), 1:92.

16.   Jefferson to Jones, January 2, 1814, 7:101. Here Jefferson added what was surely his highest accolade for Washington: "He was no monarchist from preference of his judgment. The soundness of that gave him correct views of the rights of man, and his severe justice devoted him to them." Ibid., 7:102.

17.   A. Adams to J. Adams, July 16, 1775, AFC, 1:246.

18.   Rush to Rustin, October 29, 1775, 1:92.

19.   John Adams, *Diary*, in Congress May–June 1776, DAJA, 3:336.

20.   George Washington to the Officers of Five Virginia Independent Companies, June 20, 1775, PGW-RWS, 1:16–17.

21.   Silas Deane to Elizabeth Deane, June 16, 1775, in Smith, ed., *Letters of the Delegates,* 1:494.

22.   JCC, June 16, 1775, 2:92.

23.   John Adams to Elbridge Gerry, June 18, 1775, PJA, 3:26.

24.   E.g., "London, August 5," *Pennsylvania Evening Post*, October 14, 1775, 467.

25. George Washington, "Address to the New York Provincial Assembly,"
     June 26, 1775, in PGW-RWS, 1:41.

26. Thomas Jefferson to George Washington, April 16, 1784, PTJ, 7:106–7.

27. Upon Washington's appointment in June 1775, Connecticut delegate
     Silas Deane predicted to his wife about Washington, "Our youth
     look up to This Man as a pattern to form themselves by." S. Deane to
     E. Deane, June 16, 1775, 1:494.

28. JCC, July 6, 1775, 2:154–56.

29. Benjamin Franklin to Joseph Priestley, October 3, 1775, PBF, 22:218.
     For an extract published in London from a similar letter to English
     correspondents, see Benjamin Franklin to a Friend in London, October
     3, 1775, PBF, 22:215–16.

30. George Washington to Samuel Washington, July 20, 1775, PGW-RWS,
     1:135. British general Henry Clinton made an eerily similar observation
     using almost precisely the same words. The two generals, Washington
     and Clinton, later led their respective armies when a single fateful bat-
     tle at Yorktown lost the war for Britain.

31. Benjamin Franklin to Joseph Priestley, May 16, 1775, PBF, 22:44.

32. For example, in June 1775, Franklin wrote, "Hostilities being com-
     menced by General Gage against America, and a Civil War begun,
     which I have no Chance of living to see the End of, being 70 Years of
     Age." Benjamin Franklin to Thomas Life, June 5, 1775, PBF, 22:59.

33. Benjamin Franklin to Jonathan Shipley, September 13, 1775, PBF,
     22:200; Benjamin Franklin to Joseph Priestley, July 7, 1775, PBF, 22:92.
     In the latter letter, he elaborated, "In the morning at 6, I am at the
     committee of safety, appointed by the assembly to put the province in
     a state of defense; which committee holds till near 9, when I am at the
     congress, and that sits till after 4 in the afternoon." In the former letter,
     he added, "This Bustle is unsuitable to [my] Age."

34. John Adams to Abigail Adams, July 23, 1775, AFC, 1:253.

35. E.g., B. Franklin to Life, June 5, 1775, 22:59 ("Hostilities being com-
     menced by General Gage against America," not Massachusetts); Ben-
     jamin Franklin to Humphry Marshall, May 23, 1775, PBF, 22:50–51
     ("But, as Britain has begun to use force, it seems absolutely necessary
     that we should be prepared to repel force by force, which I think,
     united, we are well able to do").

36. Benjamin Franklin, "Proposed Articles of Confederation," [before
     July 21, 1775], in PBF, 22:123.

37. Benjamin Franklin to David Hartley, May 8, 1775, PBF, 22:34.

38. George Washington to John Hancock, September 21, 1775, PGW-RWS, 2:25, 29.

39. George Washington, General Orders, July 4, 1775, in ibid., 1:54.

40. George Washington to Lund Washington, August 20, 1776, ibid., 1:335–36.

41. Benjamin Franklin to John Waring, December 17, 1763, PBF, 10:395. In response to a query from the Enlightenment era French philosophe the Marquis de Condorcet, Franklin wrote in 1774, "The Negroes who are free live among the White People, but are generally improvident and poor. I think they are not deficient in natural Understanding, but they have not the Advantage of Education." Benjamin Franklin to Marquis de Condorcet, March 20, 1774, PBF, 21:151.

42. "Minutes of the Conference Between a Committee of Congress, Washington, and Representatives of the New England Colonies," October 23, 1775, in PBF, 22:237 (among the matters that Washington raised with the delegates after the conference adjourned, with the notation that the recommendation came from the Council of Officers, which was headed by Washington). From the time of his arrival at the front, Washington complained about the "Number of Boys, Deserters, & Negroes" in the Massachusetts regiments. George Washington to [John Hancock], July 10, 1775, PGW-RWS, 1:90. The minutes of the council of officers, chaired by Washington, considering this matter state with respect to enlisting Negros in the new army, "Agreed unanimously to reject all Slaves, & by a great Majority to reject Negroes altogether." Council of War, October 8, 1775, in PGW-RWS, 2:125. The following month, recruiters were instructed to exclude Blacks.

43. George Washington, "Address to the Inhabitants of Canada," September 14, 1775, in PGW-RWS, 1:461.

44. Washington viewed Franklin's mission as "essential" to the American effort in Canada. See George Washington to Benjamin Franklin, May 20, 1776, ibid., 4:345.

45. As Franklin was preparing to head north to Canada, Washington wrote from the front about support for independence, "I find common sense is working a powerful change there in the Minds of many Men." George Washington to Joseph Reed, April 1, 1776, ibid., 4:11. Only two years earlier, Paine had moved to Philadelphia from England carrying a letter of recommendation from Franklin.

46. Benjamin Franklin to Josiah Quincy Sr., April 15, 1776, PBF, 22:400.

47. Benjamin Franklin, Samuel Chase, and Charles Carroll to John Hancock, May 1, 1776, ibid., 22:413.

48. The Commissioners in Canada to [John Hancock], May 8, 1776, ibid., 22:425.

49. Washington, "Address to the Inhabitants of Canada," 1:462.

50. "The Committee of Secret Correspondence: A Report to Congress," February 14, 1776, in PBF, 22:352.

51. "Instructions and Commission from Congress to Franklin, Charles Carroll, and Samuel Chase for the Canadian Mission," March 20, 1776, in ibid., 22:381–83.

52. Ibid., 22:382.

53. Ibid., 22:381.

54. Washington, "Address to the Inhabitants of Canada," 1:462.

55. "Instructions and Commission from Congress," 22:383.

56. George Washington to Benjamin Franklin, May 20, 1776, PBF, 22:438.

57. Jane Mecom to Catharine Greene, June 1, 1776, ibid., 22:442.

58. Benjamin Franklin to Charles Carroll and Samuel Chase, May 27, 1776, ibid., 22:440.

59. Benjamin Franklin to George Washington, June 21, 1776, ibid., 22:484.

60. George Washington to John Augustine Washington, May 31–June 4, 1776, PGW-RWS, 4:412.

61. Only limited documentation from the drafting process survives and no certain proof of who made what edit. Some handwritten drafts exist with various insertions and deletions by different pens, suggesting that these drafts were passed around and subject to committee review. Letters from the time and later recollections suggest that, after Jefferson wrote his original draft, he gave it separately to committee members Franklin and Adams for review, and also to the full committee. Even the substantial alterations by Congress cannot be fully separated from earlier edits by committee members. Historians have endlessly analyzed the available evidence and reached different conclusions on the evolution of this historic document; here I have taken a middle course, relying in part on the sources and text in "The Declaration of Independence," in PTJ, 1:413–33. Some of the edits that I ascribe to Franklin are also attributed to him in H. W. Brands, *The First American: The Life and Times of Benjamin Franklin* (New York: Doubleday, 2000), 511. Walter Isaacson swears by the "self-evident" edit in his *Benjamin Franklin: An American Life* (New York: Simon & Schuster, 2003), 312, although the eminent Jefferson scholar Julian Boyd disputes it in his *Declaration of*

Independence: The Evolution of the Text (Princeton, N.J.: Princeton University Press, 1945), 22–23.

62. Thomas Jefferson, "Biographical Sketches of Distinguished Men," in *The Writing of Thomas Jefferson,* ed. Albert Ellery Bergh (Washington, D.C.: Thomas Jefferson Memorial Association, 1907), 18:169–70.

63. George Washington, General Orders, July 9, 1776, in PGW-RWS, 5:246.

64. First published as an "anecdote" in Jared Sparks, ed., *The Works of Benjamin Franklin* (1836–1840; repr., Boston: Whittemore, 1854), 1:408.

65. Compare Henry Knox to Lucy Knox, July 8, 1776, Gilder Lehrman Institute of American History digital collection, www.gilderlehrman .org/sites/default/files/inline-pdfs/t-02437-00363.pdf, with George Washington to John Hancock, July 4–5, 1776, PGW-RWS, 5:200.

66. G. Washington to J. A. Washington, May 31–June 4, 1776, 4:413.

67. Lord Howe to George Washington, July 13, 1776, PGW-RWS, 5:296 (referring to the massive British invasion, Lord Howe begins his letter to Washington, "The Situation in which you are placed . . ."); Lord Howe to Benjamin Franklin, July 12, 1776, PBF, 22:484.

68. Benjamin Franklin to Lord Howe, July 20, 1776, PBF, 22:520.

69. Lord Howe to Benjamin Franklin, August 16, 1776, ibid., 22:565.

70. Benjamin Franklin to Lord Howe, August 20, 1776, ibid., 22:575.

71. John Sullivan, "Report of Message from Lord Howe to Congress," September 3, 1776, in JCC, 5:731.

72. John Adams, *Diary,* September 9, 1776, in DAJA, 3:418.

73. The Committee of Conference, "Report to Congress," September 17, 1776, in PBF, 22:607.

74. John Adams, *Diary,* September 17, 1776, in DAJA, 3:422.

75. Ambrose Serle, September 13, 1776, in *The American Journal of Ambrose Serle, Secretary to Lord Howe, 1776–1778,* ed. Edward H. Tatum Jr. (San Marino, Calif.: Huntington Library, 1940), 101.

76. G. Washington to J. A. Washington, May 31–June 4, 1776, 4:412.

77. Benjamin Franklin to Richard and Sarah Franklin Bache, May 10, 1785, in *The Writings of Benjamin Franklin,* ed. Albert Henry Smyth (New York: Haskell, 1970), 9:327. Writing to Congress upon his arrival in France, Franklin described the crossing as "a short but rough Passage of 30 Days." Benjamin Franklin to Committee of Secret Correspondence, December 8, 1776, PBF, 23:30. On the same day, Franklin wrote to his sister, "I arrived here safe after a Passage of 30 Days, some what fatigued and weakned by the Voyage, which was a rough one." Benjamin Franklin to Jane Mecom, December 8, 1776, PBF, 23:33.

78. Benjamin Franklin to Sarah Bache, June 3, 1779, PBF, 29:613.

79. Benjamin Franklin to Emma Thompson, February 8, 1777, ibid., 23:298.

80. The American Commissioners to Vergennes, January 5, 1777, ibid., 23:123–24.

81. "The American Commissioners to the Committee of Secret Correspondence," March 12, 1777, in ibid., 23:473.

82. Bernard Bailyn, *To Begin the World Anew: The Genius and the Ambiguities of the American Founders* (New York: Knopf, 2003), 67.

83. George Washington to John Hancock, December 9, 1776, PGW-RWS, 7:284.

84. George Washington to John Hancock, December 20, 1776, ibid., 7:382.

85. George Washington to Samuel Washington, December 18, 1776, ibid., 7:370.

86. George Washington to John Parke Custis, January 22, 1777, ibid., 8:123.

87. G. Washington to Hancock, December 20, 1776, 7:382.

88. Ibid.

89. Resolution, JCC, December 12, 1776, 6:1027 (*"Resolved* . . . that, until the Congress shall otherwise order, General Washington be possessed of full power to order and direct all things relative to the department, and to the operation of war").

90. John Adams to Abigail Adams, June 18, 1777, AFC, 2:268.

91. John Adams to Abigail Adams, July 30, 1777, ibid., 2:297. Eleven days later, Washington wrote to his brother about Howe and his army, "We have remain'd in the most perfect ignorance, and disagreeable State of Suspence, respecting their designs," and spoke of marching and countermarching his own army in response to Howe's supposed movements. George Washington to Samuel Washington, August 10, 1777, PGW-RWS, 10:581.

92. Joseph Kendall to [George] Frend, September 7, 1778, in *Letters on the American Revolution in the Library at "Karolfred,"* ed. Frederic R. Kirkland (New York: Coward-McCann, 1952), 2:44.

93. This account comes from Franklin's longtime friend Benjamin Rush, who reported Franklin saying that he "wore it to give a little revenge. I wore this Coat on the day Widderburn abused at Whitehall." Benjamin Rush, "Excerpts from the Papers of Dr. Benjamin Rush," *Pennsylvania Magazine of History and Biography* 29 (1905): 28. Franklin's recent biographers differ in their telling of this story. Presumably relying on reports from the Privy Council grilling, Gordon Wood depicts Franklin's garment as "an old blue velvet coat." Wood, *The Americanization*

of Benjamin Franklin (New York: Penguin Press, 2004), 191. Citing secondary sources, Walter Isaacson states that Franklin "dressed in a plain brown suit." Isaacson, *Franklin*, 348. Others omit the episode. E.g., Brands, *First American*, 544.

94. George Washington to Benjamin Franklin, October 9, 1780, PBF, 33:399.

95. Benjamin Franklin to George Washington, April 2, 1782, ibid., 37:89.

96. Benjamin Franklin to Barbeu-Dubourg, [after October 2, 1777], ibid., 25:21.

97. E.g., see Benjamin Franklin to George Washington, April 2, 1777, ibid., 24:551 ("I refuse every day Numbers of Applications for Letters in favour of Officers who would go to America, as I know you must have more upon your Hands already than you can well employ; but M. Turgot's Judgement of Men has great Weight").

98. Benjamin Franklin to George Washington, August 24, 1777, ibid., 24:459.

99. Benjamin Franklin, "Model of a Letter of Recommendation of a Person You Are Unacquainted With," April 2, 1777, in ibid., 24:549–50.

100. Benjamin Franklin to George Washington, June 13, 1777, ibid., 24:158.

101. Benjamin Franklin to George Washington, March 5, 1780, ibid., 32:57. Franklin sent a similar compliment a year earlier, when he asked his daughter in Philadelphia, "If you happen again to see General Washington, assure him of my very great respect, and tell him that all the old Generals here amuse themselves in studying the accounts of his operations, and approve highly of his conduct." Benjamin Franklin to Sarah Bache, June 3, 1779, ibid., 29:615.

102. George Washington to Benjamin Franklin, October 22, 1781, ibid., 35:637.

103. Benjamin Franklin to George Washington, September 20, 1785, PGW-CS, 3:267.

Chapter Five: "The Most Awful Crisis"

1. Walter Isaacson, *Benjamin Franklin: An American Life* (New York: Simon & Schuster, 2003), 349.

2. Leonidas, "On the French Alliance," *Pennsylvania Packet,* August 24, 1779, 1. When Franklin's sister Jane, who was then living in Pennsylvania, learned of the alliance, she wrote to her brother in France, "I hope now we may be restored to Peace on our own Ecqutable terms

on Established Independance." Of course, she credited and congratu-
lated her brother. Jane Mecom to Benjamin Franklin, May 5, 1778, PBF,
26:402.

3. Massachusetts Board of War to Benjamin Franklin, May 8, 1778, ibid.,
    26:420.

4. Benjamin Franklin to Massachusetts Board of War, February 17, 1778,
    ibid., 25:684.

5. George Washington, General Orders, May 5, 1778, in PGW-RWS,
    15:38–39.

6. John Adams to Benjamin Rush, April 4, 1790, PJA, --:---.

7. George Washington to Marquis de Lafayette, March 18, 1780, PGW-
    RWS, 25:83.

8. Articles of Confederation, Art. III (1781).

9. Von Steuben's companion was Peter Stephen Du Ponceau, age seven-
    teen. While at Valley Forge, von Steuben began lasting relationships
    with two young American army officers, Benjamin Walker and William
    North, who served as his aides-de-camp, were adopted as his sons, and
    became his heirs. Later, von Steuben lived with a younger man, John W.
    Mulligan Jr., who had previously lived with John Adams's son Charles
    before the elder Adams ended the arrangement. Intimate letters exist
    between the men, but not explicit descriptions of their relationships.

10. George Washington, General Orders, September 26, 1780, in WGW,
    20:95.

11. Benjamin Franklin to Marquis de Lafayette, December 9, 1780, PBF,
    34:143.

12. Washington, General Orders, September 26, 1780, 20:95.

13. George Washington to Benjamin Harrison, December 18–30, 1778,
    PGW-RWS, 18:248, 250. Dozens of letters by Washington from 1778–
    1779 express similar sentiments.

14. George Washington to Joseph Reed, May 28, 1780, WGW, 18:436.

15. George Washington to Joseph Jones, May 31, 1780, ibid., 18:453.

16. George Washington to Fielding Lewis Sr., July 6, 1780, ibid., 19:132.

17. G. Washington to Jones, May 31, 1780, 18:453.

18. G. Washington to Lewis, July 6, 1780, 19:132.

19. In one typical outburst, Washington called war profiteers, which in-
    cluded people who monopolized products and those who forestalled
    delivery to create a higher price, "pests of Society, & the greatest en-
    emies we have, to the happiness of America," and urged the states to

outlaw their actions and hang them in punishment. George Washington to Joseph Reed, December 12, 1778, PGW-RWS, 18:397.

20. George Washington to Samuel Huntington, January 23, 1781, WGW, 21:136.

21. John Adams, *Diary*, May 27, 1778, in DAJA, 4:118.

22. Benjamin Franklin to Robert R. Livingston, July 22, 1783, PBF, 40:358. Regarding Lee, Franklin wrote, "I take no revenge of such enemies, than to let them remain in the miserable situation in which their malignant natures have placed them." Benjamin Franklin to Richard Bache, June 2, 1779, ibid., 29:599.

23. Gordon S. Wood, *The Americanization of Benjamin Franklin* (New York: Penguin Press, 2004), 196.

24. Benjamin Franklin to Alexander Gillon, July 5, 1779, PBF, 30:37.

25. George Washington to John Laurens, April 9, 1781, WGW, 21:438.

26. Even Washington, in his letter to Laurens, spoke of *"the impracticality of carrying on the War without"* the funds that Franklin was directed to seek from France. Ibid (emphasis in original).

27. Benjamin Franklin to Samuel Huntington, March 12, 1781, PBF, 34:446.

28. Benjamin Franklin to William Carmichael, August 24, 1781, ibid., 35:399.

29. Robert R. Livingston to Benjamin Franklin, October 20, 1781, ibid., 35:618.

30. Benjamin Franklin to Comte de Vergennes, November 20, 1781, ibid., 36:79.

31. George III to Lord Shelburne, September 16, 1782, in *Boswell's Life of Johnson,* ed. George Birkbeck Hill (Oxford: Clarendon Press, 1887), 3:241n2.

32. Benjamin Franklin, "Journal of the Peace Negotiations," May 9, 1782, in PBF, 37:291.

33. George Washington to Benjamin Lincoln, October 2, 1782, WGW, 25:227–28.

34. Henry Knox et al., "To the United States in Congress Assembled," December 1782, in JCC, 24:291.

35. For example, at the time, the federalist-minded congressman James Madison wrote that officers' petition would "furnish new topics in favor the Impost." James Madison to Edmund Randolph, December 30, 1782, PJM, 5:473.

36. Gouverneur Morris to John Jay, January 1, 1783, in Jared Sparks, *The Life of Gouverneur Morris, with Selections from His Correspondence* (Boston: Gray & Bowen, 1832), 1:249.

37. Alexander Hamilton to George Washington, February 13, 1783, PAH, 3:254. Although opposed to using the army to secure it, Washington depicted himself as "a warm friend to the Impost." E.g., George Washington to William Gordon, July 8, 1783, WGW, 27:49.

38. Ron Chernow, the biographer of Washington and Hamilton, wrote about the incident, "In suggesting that Washington exploit the situation to influence Congress, Hamilton toyed with combustible chemicals." Ron Chernow, *Washington: A Life* (New York: Penguin Press, 2010), 433. At the same time and probably in league with Hamilton, Gouverneur Morris sent a similar letter to Nathanael Greene, the leader of American forces in the south, suggesting that the army would be paid only if it united in demanding it. Richard Brookhiser, *Gentleman Revolutionary: Gouverneur Morris— The Rake Who Wrote the Constitution* (New York: Free Press, 2003), 72.

39. George Washington to Alexander Hamilton, March 4, 1783, WGW, 26:186–87.

40. Along with Gates, Washington at first blamed Robert Morris but then shifted his accusation to Gouverneur Morris (no relation), who then served as Robert Morris's assistant. Compare George Washington to Alexander Hamilton, April 4, 1783, WGW, 26:293, with George Washington to Alexander Hamilton, April 16, 1783, ibid., 26:324. Colonel Walter Stewart, then living in Philadelphia and serving at Washington's request as inspector general of the army's northern department, carried letters from Robert and Gouverneur Morris to Newburgh, where Stewart met with Gates and others. See, generally, Charles Rappleye, *Robert Morris: Financier of the Revolution* (New York: Simon & Schuster, 2010), 331–51; and Richard H. Kohn, "The Inside History of the Newburgh Conspiracy," *William and Mary Quarterly* 27, no. 2 (1970): 205–6.

41. See James Madison, "Notes of Debates," February 20, 1783, in JCC, 25:906–7. Also see, generally, John Ferling, *The Ascent of George Washington* (New York: Bloomsbury, 2009), 231–33.

42. "To the Officers of the Army," [March 10, 1783], in JCC, 24:297.

43. George Washington, General Orders, March 11, 1783, in WGW, 26:208.

44. George Washington, "To Officers of the Army," March 15, 1783, in ibid., 26:226–27.

45. Ibid., 26:222n38.

46. Horatio Gates, [Minutes of Meeting of Officers], March 15, 1783, in JCC 24:311.

47. Benjamin Franklin to Robert R. Livingston, April 15, 1783, PBF, 39:472 ("The Finances here are embarrass'd, & a new Loan is proposed by

way of Lottery, in which it is said by some Caculators, the King will pay at the Rate of 7 per. Cent. I mention this to furnish you with a fresh convincing Proof, against Cavillers of the Kings Generosity towards us, in lending us six Millions this Year at 5 per Cent. and of his Concern for our Credit, in saving by that Sum the honour of Mr. Morris's Bills, while those drawn by his own Officers abroad, have their Payment suspended for a Year after they become due").

48. G. Washington to Hamilton, April 4, 1783, 26:293.

49. See George Washington to Theordorick Bland, April 4, 1783, WGW, 26:288 (first letter).

50. George Washington, "Sentiments on a Peace Establishment," May 1, 1783, in ibid., 26:375.

51. George Washington to the States (Circular), June 8, 1783, ibid., 26:485–89.

52. E.g., George Washington to Marquis de Lafayette, August 15, 1786, PGW-CS, 4:215 (tellingly referring to himself as one "member of an infant-empire").

53. G. Washington to the States, June 8, 1783, 26:496–97.

54. James Thomas Flexner, *George Washington in the American Revolution* (Boston: Little, Brown, 1967), 514.

55. E.g., G. Washington to Gordon, July 8, 1783, 27:49.

56. George Washington to Marquis de Lafayette, April 5, 1783, WGW, 26:298.

57. George Washington to John Augustine Washington, June 15, 1783, ibid., 27:12.

58. George Washington, "Farewell to the Armies of the United States," November 2, 1783, in ibid., 27:222–27.

59. George Washington, Address, December 23, 1783, in JCC, 25:838.

60. Samuel Huntington to Benjamin Franklin, June 19, 1781, PBF, 35:175.

61. Benjamin Franklin to Robert R. Livingston, December 5–14, 1782, ibid., 38:416–17.

62. Edmund S. Morgan, *Benjamin Franklin* (New Haven: Yale University Press, 2002), 304.

63. Robert Morris to Benjamin Franklin, September 27, 1782, PBF, 38:147.

64. Benjamin Franklin to Robert Morris, December 25, 1783, ibid., 41:347–48 (a later response and maybe tempered by time but reflecting Franklin's ongoing views).

65. Benjamin Franklin to Charles Thomson, May 13, 1784, ibid., 42:244 (Thomson was Congress's longtime secretary).

66. Benjamin Franklin to William Straham, August 19, 1784, ibid., 43:29. Immediately after this sentence, Franklin added, "If I had ever before been an Atheist I should now have been convinced of the Being and Government of a Deity. It is he who abases the Proud and favors the Humble!"

67. Benjamin Franklin to Abbés Chalut and Arnoux, April 17, 1787, ibid., 46:---.

68. Benjamin Franklin to Henry Laurens, March 12, 1784, ibid., 42:49. In a postscript, Franklin added his "Thanks for your kind Assurances of never forsaking my Defence should there be need. I apprehend that the violent Antipathy of a certain Person to me may have produc'd some Calumnies, which what you have seen and heard here may enable you easily to refute. You will thereby exceedingly oblige one, who has liv'd beyond all other Ambition than that of dying with the fair Character he has long endeavour'd to deserve."

69. Benjamin Franklin to John and Sarah Jay, May 13, 1784, ibid., 42:242.

70. John Jay to Benjamin Franklin, March 8, 1785, ibid., 43:475–76.

71. Comte de Vergennes to François Barbé de Marbois, May 10, 1785, in *The Emerging Nation: A Documentary History of the Foreign Relations of the United States under the Articles of Confederation, 1780–1789,* ed. Mary A. Giunta et al. (Washington, D.C.: National Historical Publications, 1996), 2:626.

72. When Congress tapped Jefferson to serve as its next ambassador to France, he stressed that he would simply succeed Franklin—no one could replace him, Jefferson said—and he later hailed Franklin as "the ornament of our country and I may say of the world." Thomas Jefferson to Ferdinand Grand, April 23, 1790, PTJ, 16:369.

73. Francis Hopkinson to Benjamin Franklin, May 24, 1784, PBF, 42:272.

74. George Washington to Benjamin Franklin, September 25, 1785, PGW-CS, 3:275n1.

75. George Washington to Armand-Louis de Gontaut, duc de Lauzun, February 1, 1784, ibid., 1:91. See also George Washington to Robert Morris, January 4, 1784, ibid., 1:11 ("this retreat from my public cares").

76. George Washington to Charles Bennett, fourth earl of Tankerville, January 20, 1784, ibid., 1:65.

77. See Edmund S. Morgan, "George Washington: The Aloof American," in *George Washington Reconsidered,* ed. Don Higginbotham (Charlottesville: University of Virginia Press, 2001), 290–91.

78. See Henry Wiencek, *An Imperfect God: George Washington, His Slaves, and the Creation of America* (New York: Farrar, Straus and Giroux, 2003), 92–96.

79. George Washington to Fielding Lewis Jr., February 27, 1784, PGW-CS, 1:161.

80. George Washington to Jonathan Trumbull Jr., January 5, 1784, ibid., 1:12.

81. George Washington to Benjamin Harrison, January 18, 1784, ibid., 1:57.

82. George Washington to Thomas Jefferson, March 29, 1784, ibid., 1:239.

83. George Washington to Benjamin Harrison, October 10, 1784, ibid., 2:92.

84. George Washington to Richard Henry Lee, June 18, 1786, ibid., 4:118.

85. George Washington to Jacob Read, November 3, 1784, ibid., 2:121.

86. G. Washington to Harrison, October 10, 1784, 2:92.

87. George Washington to Benjamin Franklin, September 26, 1785, PGW-CS, 3:275.

88. George Washington to James Madison, November 30, 1785, ibid., 3:420.

89. Annapolis Convention, "Address of the Annapolis Convention," September 14, 1786, in PAH, 3:687.

90. James Madison to Thomas Jefferson, August 12, 1786, PTJ, 10:233.

91. Alexander Hamilton, "Address," September 14, 1786, in PAH, 3:689.

92. Benjamin Franklin to Commissioners, October 26, 1786, PBF, 45:---.

93. George Washington to Comte de Rochambeau, December 1, 1785, PGW-CS, 3:427.

94. G. Washington to B. Franklin, September 26, 1785, 3:275.

95. E.g., compare George Washington to Marquis de Lafayette, December 8, 1784, PGW-CS, 2:175 (speaking of his expectation of soon being "entombed in the dreary mansions of my fathers") with Benjamin Franklin to George Whately, August 21, 1784, PBF, 43:44 ("I look upon death to be as necessary to our constitutions as sleep. We shall rise refreshed in the morning."). These contrasting letters—one fatalistic; one optimistic—were written within months of each other.

96. G. Washington to Rochambeau, December 1, 1785, 3:427.

97. Benjamin Franklin to Jane Mecom, September 19, 1785, PBF, 44:---. In this letter, he adds about his homecoming, "I am continually surrounded by congratulating Friends."

98. George Washington to Marquis de Lafayette, November 8, 1785, PGW-CS, 3:345. At the time, Franklin explained his willingness to serve in a letter to Thomas Paine as follows "For my Fellow Citizens having in a considerable Body express'd their Desire that I would still

take a Post in their publick Councils, assuring me it was the unanimous Wish of the different Parties that divide the State, from an Opinion that I might find some means of reconciling them; had not sufficient Firmness to refuse their Request." Benjamin Franklin to Thomas Paine, September 27, 1785, PBF, 44:---.

99. G. Washington to Rochambeau, December 1, 1785, 3:428. Expressing similar sentiments, Jay wrote to Franklin, "It strikes me that you will find it somewhat difficult to manage the two Parties in Pensylvania. It is much to be wished that union and Harmony may be re-established there, and if you accomplish it, much Honor and many Blessings will result from it. Unless you do it, I do not know who can." John Jay to Benjamin Franklin, October 4, 1785, PBF, 44:---.

100. Benjamin Rush to Richard Price, May 25, 1786, in *Letters of Benjamin Rush*, ed. L. H. Butterfield (Princeton, N.J.: Princeton University Press, 1951), 1:389–90.

101. Perhaps reflecting on reasons for the success of this alternative currency scheme and his own impending mortality, Franklin soon coined the enduring aphorism "In this world nothing can be said to be certain, except death and taxes." Benjamin Franklin to M. Le Roy, November 13, 1789, in Jared Sparks, ed., *The Works of Benjamin Franklin* (Boston: Hilliard, 1840), 10:410.

102. Benjamin Franklin to Thomas Jefferson, March 20, 1786, PTJ, 9:349; Benjamin Franklin to Thomas Jefferson, April 19, 1787, ibid., 11:302. In a similar vein and about the same time as the earlier of these two letters to Jefferson, Franklin told a correspondent in France, in a letter that understandably downplayed but nevertheless admitted Franklin's concerns, "There are some few Faults in our Constitutions, which is no wonder, considering the stormy Season in which they were made, but these will soon be corrected." Benjamin Franklin to Rodolphe-Ferdinand Grand, March 5, 1786, PBF, 45:---. Similarly, in July, Franklin wrote to a correspondent in England about government in America, "Notwithstanding some political Errors [we have] to eradicate, I flatter my self that on the whole and in time we shall do very well." Benjamin Franklin to Richard Price, July 29, 1786, ibid., 45:---. At the Constitutional Convention, Franklin consistently supported giving broad powers to the federal government and offered proposals to expand them.

103. Benjamin Franklin to Edward Bancroft, November 26, 1786, PBF, 45:---. Bancroft was an American-born English loyalist who served as the secretary to the American delegation in Paris while spying for Britain.

104. Edmund Randolph to Benjamin Franklin, December 1, 1786, ibid., 45:---
(first letter); Edmund Randolph to Benjamin Franklin, December 1,
1786, ibid., 45:--- (second letter; emphasis added); Edmund Randolph to
Benjamin Franklin, December 6, 1786, ibid., 45:---.

105. Benjamin Franklin to Edmund Randolph, December 21, 1786, ibid.,
45:---.

106. Benjamin Franklin to Marquis de Chastellux, April 17, 1787, ibid.,
46:--- ("assembly of Notables"); Benjamin Franklin to Thomas Jefferson, April 19, 1787, ibid., 46:--- ("If it does not do Good").

107. As the delegates assembled in May, Franklin called their work "[a]
most important business," and expressed his "hope" that it "will be
attended with Success." Benjamin Franklin to Richard Price, May 18,
1787, ibid., 46:---.

108. Erasmus Darwin to Benjamin Franklin, May 29, 1787, ibid., 46:---.

Chapter Six: Rendezvous in Philadelphia

1. Benjamin Franklin to Jane Mecom, September 21, 1786, PBF, 45:---.

2. Benjamin Franklin to Louis-Guillaume Le Veillard, April 17, 1787,
PBF, 46:---.

3. Ibid.

4. Benjamin Franklin to Jane Mecom, May 30, 1787, PBF, 46:---.

5. George Fox to George Washington, May 14, 1787, PGW-CfS, 5:187.

6. Jane Mecom to Benjamin Franklin, May 22, 1787, PBF, 46:---.

7. Benjamin Franklin to George Washington, April 3, 1787, PGW-CfS,
5:122.

8. Robert Morris to George Washington, April 23, 1787, ibid., 5:153.

9. Charles Biddle to Benjamin Franklin, May 1787, PBF, 46:---.

10. George Washington to Edmund Randolph, December 21, 1786, PGW-CfS, 4:472.

11. Edmund Randolph to George Washington, January 4, 1787, ibid., 4:501.

12. James Madison to George Washington, December 24, 1786, PJM, 9:224.

13. George Washington to James Madison, March 31, 1787, PGW-CfS, 5:116.

14. Resolution, February 21, 1787, JCC, 32:74.

15. Henry Knox to George Washington, April 9, 1787, PGW-CfS, 5:133–34.

16. Henry Knox to George Washington, January 14, 1787, ibid., 4:522.
Knox went on to explain, "The laws passed by the general governmt
to be obeyed by the local governments, and if necessary to be enforced
by a body of armed men to be kept for the purposes which should be
designated—All national objects, to be designed and executed by the

general government, without any reference to the local governments. This rude sketch is considered as the government of the least possible powers, to preserve the confederated government—To attempt to establish less, will be to hazard the existence of republicanism, and to subject us, either to a division of the European powers, or to a despotism arising from highhanded commotions."

17. John Jay to George Washington, January 7, 1787, ibid., 4:503. Jay preceded this comment with the observation about the general government, "What Powers should be granted to the Government so constituted is a Question which deserves much Thought—I think the more the better."

18. James Madison to George Washington, April 16, 1787, ibid., 5:145. In this letter, Madison added, "I would propose next that in addition to the present federal powers, the national Government should be armed with positive and compleat authority in all cases which require uniformity; such as the regulation of trade, including the right of taxing both exports & imports, the fixing the terms and forms of naturalization &c. &c."

19. George Washington to Edmund Randolph, March 28, 1787, ibid., 5:113. In particular, Washington wrote, "As my friends, with a degree of sollicitude which is unusual, seem to wish my attendance on this occasion, I have come to a resolution to go if my health will permit."

20. George Washington to John Jay, March 10, 1787, ibid., 5:79–80.

21. Henry Knox to George Washington, April 19, 1787, ibid., 5:96. Knox here wrote, "Were the convention to propose only amendments, and patch work to the present defective confederation, your reputation would in a degree suffer—But were an energetic, and judicious system to be proposed with Your signature, it would be a circumstance highly honorable to your fame, in the judgement of the present and future ages."

22. B. Franklin to G. Washington, April 3, 1787, 5:122.

23. Henry Knox to George Washington, October 23, 1786, PGW-CfS, 4:300–301.

24. George Washington to Henry Knox, December 26, 1786, ibid., 4:481.

25. Ibid.

26. George Washington to Henry Knox, February 3, 1787, ibid., 5:7–8.

27. Benjamin Franklin to James Bowdoin, March 6, 1787, PBF, 46:---.

28. George Washington to Marquis de Lafayette, March 25, 1787, PGW-CfS, 5:106.

29. George Washington, "Journal," May 9, 1787, in DGW, 5:153.

30. George Washington to Robert Morris, May 5, 1787, PGW-CfS, 5:171.

31. For a discussion of Franklin's self-expressed "middling" status, see Gordon S. Wood, *The Americanization of Benjamin Franklin* (New York: Penguin Press, 2004), 46–49.

32. George Washington, "Journal," May 13, 1787, in DGW, 5:155.

33. Ibid. Washington depicted his visit to Franklin as occurring "as soon as I got to Town."

34. Richard Beeman, *Plain, Honest Men: The Making of the American Constitution* (New York: Random House, 2009), 35–36.

35. Benjamin Franklin to Thomas Jordan, May 18, 1787, PBF, 46:---.

36. Benjamin Franklin to Louis-Guillaume Le Veillard, April 15, 1787, ibid., 46:---.

37. William Pierce, "Character Sketches of Delegates to the Federal Convention," in Farrand, 3:91.

38. Benjamin Franklin to Abbés Chalut and Arnoux, April 17, 1787, PBF, 46:---.

39. Benjamin Franklin to Thomas Jefferson, April 19, 1787, ibid., 46:---.

40. John Adams to Benjamin Franklin, January 27, 1787, ibid., 46:--- (in which Adams seemed to anticipate Franklin's response to the book by stating, "It contains my Confession of political Faith and if it be Heresy, I shall I Suppose be cast out of Communion. But it is the only Sense, in which I am or ever was a Republican"); Benjamin Franklin to John Adams, May 18, 1787, ibid., 46:---.

41. Beeman, *Plain, Honest Men*, 53.

42. Catherine Drinker Bowen, *Miracle at Philadelphia: The Story of the Constitutional Convention, May to September, 1787* (Boston: Little, Brown, 1966), 18.

43. George Mason to George Mason Jr., May 20, 1787, PGM, 3:880.

44. George Mason to Arthur Lee, May 21, 1787, ibid., 3:882.

45. Farrand, May 31, 1787, 1:48, 546.

46. Farrand, May 25, 1787, 1:3–5.

47. See Farrand, May 25, 1787, 1:4 (Madison), 6 (Yates); Luther Martin, "Genuine Information," in ibid., 3:173.

48. George Mason to George Mason Jr., May 27, 1787, PGM, 3:884.

49. Pierce, "Character Sketches," in Farrand, 3:91.

50. Farrand, May 28, 1787, 1:15.

51. Jared Sparks, "Journal," April 19, 1830, in Farrand 3:479 (Sparks's notes on visit to James Madison).

52. Various delegates did jot down some firsthand notes about the Conven-

tion in private journals or letters that became public later but the "Notes" taken by Madison provide the closest thing to a comprehensive account of the closed-door sessions. He vowed to keep them confidential until the last delegate died, which turned out to be him. For an analysis of the accuracy of Madison's Notes, see Mary Sarah Bilder, *Madison's Hand: Revising the Constitutional Convention* (Cambridge: Harvard University Press, 2015). Among other delegates at the Convention taking notes, New York antifederalist Robert Yates kept the second most detailed ones. Maryland antifederalist Luther Martin also compiled his own account of key events. Some of these notes and accounts, most notably those by Yates and Martin, became public after the Convention ended.

53. George Washington, "Journal," May 28, 1787, in DGW, 3:220.
54. William Pierce, "Anecdote," in Farrand, 3:86–87.
55. Farrand, May 29, 1787, 1:23–24.
56. *Journals of the American Congress: From 1774 to 1788* (Washington, D.C.: Way & Gideon, 1823), 4:A39 (credentials of the members from Georgia).
57. Farrand, May 29, 1787, 1:20–21 (text of Randolph's resolutions).
58. George Washington to Thomas Jefferson, May 30, 1787, PGW-CfS, 5:208. Evidencing his support for the Virginia Plan, at the Convention Franklin praised "Mr. Randolph for having brought forth the plan in the first instance." Farrand, September 17, 1787, 2:646.
59. Benjamin Franklin to Pierre-Samuel du Pont de Nemours, June 9, 1788, PBF, 47:---.
60. Ibid. Regarding Franklin as a chess player, Jefferson recalled a time during a game Franklin played with the Duchess of Bourbon in France during the American Revolution: "Happening once to put her king into prise, the Doctr took it. 'ah, says she, we do not take kings so.' 'we do in America,' says the Doctor." Thomas Jefferson, "Anecdotes of Benjamin Franklin," December 4, 1818, in PTJ-RS, 13:463.
61. In terms of population, Connecticut at the time tied for seventh among the thirteen states, or exactly in the middle.
62. Farrand, June 30, 1787, 1:483.
63. Ibid., 1:490.
64. Ibid., 1:488, 499.
65. Ibid., 1:482, 494.
66. Ibid., 1:499.
67. George Washington to George Augustine Washington, June 3, 1787, PGW-CfS, 5:219.

68. Jared Sparks, *The Life of Gouverneur Morris, with Selections from His Correspondence* (Boston: Gray & Bowen, 1832), 1:283.

69. Farrand, July 2, 1787, 1:510–16, 519.

70. Ibid., July 5, 1787, 1:526. Franklin accepted the compromise only as a whole and objected to the Convention's considering it in parts. See ibid., July 6, 1787, 1:453.

71. Ibid., August 8, 1787, 2:224. For Mason's vehemence on this issue, see also ibid., August 13, 1787, 2:274 ("the pursestrings should be in the hands of the Representatives of the people").

72. Ibid., July 6, 1787, 1:546 (Madison reported Franklin as saying, "It was a maxim that those who feel can best judge. This end would, he thought, be best attained, if money affairs were to be confined to the immediate representatives of the people").

73. For the first vote, see ibid., August 8, 1787, 2:225; for the second vote and Madison's comment, see ibid., August 13, 1787, 2:280.

74. Ibid., September 10, 1787, 2:568.

75. Ibid., June 11, 1787, 1:192.

76. Ibid., 1:205, quotes Franklin as saying, "Representation ought to be in proportion to the importance of numbers or wealth in each state."

77. Ibid., 1:201.

78. Ibid., 1:193. They borrowed this formula from a 1783 proposal passed by Congress but never ratified by the states for equitably allocating requisitions among the states.

79. Ibid., 1:208.

80. For a clear expression of this distinction by Morris, see his comments in ibid., July 13, 1787, 1:604.

81. Ibid., July 9, 1787, 1:561.

82. Ibid., July 13, 1787, 1:605.

83. Ibid., July 9, 1787, 1:560.

84. Beeman, *Plain, Honest Men*, 207.

85. In the critical exchange on the second issue, South Carolina's Rutledge declared, "If the Convention thinks that N.C; S.C. & Georgia will ever agree to the plan, unless their right to import slaves be untouched, the expectation is in vain," to which Connecticut's Sherman conceded, "It was better to let the S. States import slaves than to part with them, if they made that a sine qua non." Farrand, August 22, 1787, 2:373–74. As for the first issue, the North Carolina delegation reported to its governor regarding the Constitution, "The Southern States have also a much better Security for the Return of Slaves who might endeavor to Escape

than they had under the original Confederation." William Blount et al. to Richard Caswell, September 28, 1787, in Farrand, 3:84.

86. Ibid., August 7, 1787, 2:204–5. In a similar vein, Franklin constantly argued for shorter residency requirements for immigrants to serve in Congress. "When foreigners after looking about for some other Country in which they can obtain more happiness, give a preference to ours, it is proof of an attachment which ought to excite our confidence & affection," he explained. Ibid., August 9, 1787, 2:249.

87. Ibid., August 10, 1787, 2:249.

88. United States Constitution, Art. I, sect. 2.

89. Farrand, September 17, 1787, 2:644. In his analysis of Washington's decision to speak on this relatively minor point after he had refrained from speaking substantively on other matters, Glenn Phelps agrees that it "reveals the pragmatic side of [Washington's] constitutional thinking." Phelps paints the amendment as a minor concession to small-scale republicans calculated to help secure ratification. Glenn A. Phelps, *George Washington and American Constitutionalism* (Lawrence: University Press of Kansas, 1993), 101.

90. Farrand, May 29 and June 1–2, 1787, 1:21, 63, 94 (quote on 21).

91. Ibid., June 1, 1787, 1:65.

92. See Max Farrand, *The Fathers of the Constitution: A Chronicle of the Establishment of the Union* (New Haven: Yale University Press, 1921), 111.

93. W.P., "An Ode," *Virginia Gazette* (Dixon and Hunter), August 24, 1776, 8.

94. George Washington to Josiah Quincy, March 24, 1776, PGW-RWS, 3:528.

95. "Anecdote," in Farrand, 3:85.

96. Farrand, June 1, 1787, 1:65.

97. Ibid., June 4, 1787, 1:103.

98. Ibid., July 26, 1787, 2:120.

99. Ibid., June 2, 1787, 1:83. Introducing this argument, Franklin depicted the love of power and the love of money as two passions that influence human affairs: "Separately each of these has great force in promoting men to action; but when united in view of the same object, they have in many minds the most violent effect," he warned (ibid., 1:82).

100. E.g., after Mason decried Morris's effort to eliminate impeachment with the rhetorical question "Shall any man be above Justice?" Franklin agreed and added with his well-known biting wit that, without impeachment, the only "recourse was . . . assassination in wch. [case the president] was deprived not only of his life but of the opportunity of vindicating his character." Ibid., July 20, 1787, 2:65.

101. Ibid., June 1, 1787, 1:66. Randolph repeated this warning the next day. Ibid., June 2, 1787, 1:72 ("A single Person may be considered the foetus of a Monarchy").

102. Ibid., June 2, 1787, 1:83.

103. Ibid., June 4, 1787, 1:102n14. Franklin went on to warn, "Foreign governments can never have that Confidence, in the Treaties or Friendship of such a Government as that which is conducted by a Number." He then asked, "The Single Head may be Sick. Who is to conduct the Public Affairs in that Case?"

104. Pierce Butler to Weedon Butler, May 5, 1788, ibid., 3:302.

105. Ibid., June 4, 1787, 1:97.

106. Benjamin Franklin Bache to Benjamin Franklin, August 1, 1787, PBF, 46:---.

107. Benjamin Franklin to Jane Mecom, September 20, 1787, ibid., 46:---.

108. Farrand, August 6, 1787, 2:185.

109. From the Convention's outset, there was a clear sense that Congress should hold power over war and peace. See ibid., June 1, 1787, 1:66–70. The Committee of Detail's draft Constitution assigned Congress the power to make war and the Senate the power to draft treaties, including peace treaties. On August 17, with little debate, the Convention changed the wording of this congressional power: "To make war" became "To declare war." Apparently the delegates saw this as a clarifying amendment. Madison and Gerry proposed it in response to Charles Pinckney's concern that vesting the power of making war in Congress might cause too much delay in repelling a sudden attack. Sherman then added, "The Executive should be able to repel and not to commence war." In support of the change, Mason said that he "was against giving the power of war to the Executive, because [he was] not safely to be trusted with it." In a similar vein, Rufus King noted "that 'make' war might be understood to 'conduct' it which was an Executive function." Ibid., August 17, 1787, 2:318–19.

110. The phrase "three fifths of all other Persons" comes from Article I, section 2, clause 3 of the final draft Constitution but, although agreed to, was not set forth in the Committee of Detail draft.

111. In addition to the shift of duties from the Senate to the presidency, which are discussed below, during the final deliberations, the power to adjudicate disputes between states passed from the Senate to the Supreme Court while the duty to try impeachments passed from the Supreme Court to the Senate. See Farrand, August 24, 1787, 2:401; ibid,

August 27, 1787, 2:427; ibid., September 4, 1787, 2:493; and ibid., September 8, 1787, 2:554.

112. Ibid., September 7, 1787, 2:542.

113. Ibid., August 8, 1787, 2:224.

114. On June 30, for example, Madison complained, "The plan in its present shape makes the Senate absolutely dependent on the States. The Senate therefore is only another edition of [the Confederation] Cong[res]s. [I know] the faults of that Body & ha[ve] used a bold language ag[ain]st it." Ibid., June 30, 1787, 1:491.

115. Ibid., September 4, 1787, 2:493–95. For early advocacy of this approach, see ibid., July 18, 1787, 2:41–44. The key vote "for Appointing the President by electors" passed the Convention by a margin of nine to two. Ibid., September 6, 1787, 2:525.

116. During the final weeks of the Convention "a growing reaction against the Senate worked in favor of the presidency," historian of the Constitution Jack Rakove notes. "The growth of the presidency owed more to doubts about the Senate than to the enthusiasm with which Hamilton, Morris, and Wilson endorsed the virtues of an energetic administration." Jack N. Rakove, *Original Meanings: Politics and Ideas in the Making of the Constitution* (New York: Knopf, 1997), 163, 167. The widely regarded republican virtues and national-mindedness of Washington (the presumptive first president) surely contributed to the attraction of the presidency over the potentially aristocratic and state-minded Senate as a repository of power. With respect to Washington, South Carolina's Pierce Butler stated that his fellow delegates "shaped their Ideas and Powers to be given to the President, by their opinions of his Virtue." P. Butler to W. Butler, May 5, 1788, 3:302.

117. The committee's recommendations authorizing the president to make treaties and nominate judges and officers with the advice and consent of the Senate passed the Convention without dissent. Farrand, September 7, 1787, 2:538–39. Rather than entrust either the executive or legislature with complete power to name judges, Franklin offered the pragmatic alternative of letting lawyers nominate judges, as done in Scotland. There, he noted, they "always select the ablest of the profession to get rid of him, and share his practice." Ibid., June 5, 1787, 1:120.

118. For example, at one point, Hamilton painted the single-term president as a "Monster" who would "consider his 7 years as 7 years of lawful plunder." Ibid., September 6, 1787, 2:524, 530.

119. Ibid., September 15, 1787, 2:632. Expanding on his objections to the presidency, Mason noted (1) the lack of an executive council, which he feared would leave the president "directed by minions and favorites," and (2) the unrestrained pardon power, "which may be sometimes exercised to screen from punishment those whom he had secretly instigated to commit the crime, and thereby prevent a discovery of his own guilt." Ibid., 2:638–39.

120. James McHenry, "Anecdotes," in Farrand, 3:85.

121. The text of the Virginia Plan is in ibid., May 29, 1787, 1:20–21.

122. George Washington to Edmund Randolph, January 8, 1788, PGW-CfS, 6:18. Washington expressed almost identical sentiments to an antifederalist former Virginia governor less than a week after the Convention ended: "I sincerely believe it is the best that could be obtained at this time." George Washington to Benjamin Harrison, September 24, 1787, ibid., 5:339.

123. George Washington to Marquis de Lafayette, February 7, 1788, ibid., 6:96 (Washington added about the proposed federal government, "No objection ought to be made against the quantity of Power delegated to it"). Franklin also never questioned the powers granted to Congress and, at the Convention, proposed increasing them by adding the power to cut canals, which opened into a general discussion of congressional power to charter corporations (including banks) that ultimately ended in leaving these powers unstated and thus perhaps implied.

124. Marquis de Lafayette to George Washington, February 4, 1788, ibid., 6:85.

125. George Washington to Marquis de Lafayette, April 28, 1788, ibid., 6:244–45. Noting how selection by independent electors guarded against "bribery and undue influence in the choice of President," Washington commented on term limits and the power of impeachment, "There cannot, in my Judgement, be the least danger that the President will by any practicable intrigue ever be able to continue himself one moment in office, much less perpetuate himself in it—but in the last stage of corrupted morals and political depravity." Of course, Washington could not foresee how the system would evolve under partisan government.

126. Farrand, September 17, 1787, 2:642. Franklin opened his speech by saying, "I confess that there are several parts of this Constitution which I do not at present approve, but I am not sure I shall never approve them." Ibid., 2:641. Echoing Franklin, Washington soon wrote, "There

are somethings in the new form, I will readily acknowledge, wch never did, and I am persuaded never will, obtain my *cordial* approbation." G. Washington to Randolph, January 8, 1788, 6:18. Addressing his remarks to Washington at the Convention, Franklin stated, "I consent, Sir, to this Constitution because I expect no better, and because I am not sure, that it is not the best." Farrand, September 17, 1787, 2:643.

127. Ibid., September 17, 1787, 2:642. Regarding the compromises, three days later, Franklin wrote about the Constitution, "The Forming of it so as to accomodate all the different Interests and Views was a difficult task." B. Franklin to Mecom, September 20, 1787, 46:---. After further reflection, Franklin observed about the need for compromise in such settings, "The wisest must agree to some unreasonable things, that reasonable ones of more consequence may be obtained." B. Franklin to du Pont de Nemours, June 9, 1788, 46:---. Echoing Franklin's view that authority conferred on the president might lead to tyranny, Washington wrote privately to Lafayette that it would happen only "in the last stage of corrupted morals and political depravity" but added, "When a people shall have become incapable of governing themselves and fit for a master, it is of little consequences from what quarter he comes." G. Washington to Lafayette, April 28, 1788, 6:245.

128. Benjamin Franklin to Count Castiglione, October 14, 1787, PBF, 46:---.

129. Regarding the novelty of a people framing their own form of government, see, e.g., George Washington to John Lathrop, June 22, 1788, PGW-CfS, 6:349 ("a new phenomenon in the political & moral world"); George Washington to Edward Newenham, August 29, 1788, ibid., 6:387–88 (a "novel & astonishing Spectacle").

130. Benjamin Franklin to Rodolphe-Ferdinand Grand, October 22, 1787, PBF, 46:---.

131. Farrand, September 17, 1787, 2:666–67 (letter to Congress, emphasis added).

132. Regarding the rapid publication of Franklin's speech, Max Farrand wrote, "Franklin seems to have sent copies of this speech in his own handwriting to several of his friends, and one of these soon found its way into print." Ibid., 2:614n1. Characterizing it as "insinuating persuasive," Maryland delegate James McHenry cynically observed that whatever happened to the Constitution, Franklin's speech "guarded the Doctor's fame." Ibid., September 17, 1787, 2:649.

133. Ibid., 2:648.

134. Ibid., 2:649.

135. Writing in his private diary on the body's closing day, Washington depicted the work of the Convention as "momentous." George Washington, "Journal," September 17, 1787, in DGW 5:185.

136. Benjamin Franklin to Jane Mecom, November 4, 1787, PBF, 46:---.

Chapter Seven: Darkness at Dawn

1. George Washington to Marquis de Lafayette, February 7, 1788, PGW-CfS, 6:95 (calling it "little short of a miracle, that the Delegates from so many different states . . . should unite in forming a system of national Government"). One noted historian would use this point in the title for her book about the Convention: Catherine Drinker Bowen, *Miracle at Philadelphia: The Story of the Constitutional Convention, May to September, 1787* (Boston: Little, Brown, 1966).

2. A Confederationalist, "To the Editor," *Pennsylvania Evening Herald,* October 27, 1787, 2.

3. "The Grand Constitution," *Columbian Herald,* November 8, 1787, 4.

4. "From a Correspondent," *Pennsylvania Packet,* September 22, 1787, 3.

5. "Assembly Proceedings," September 17, 1787, in DHRC, 2:58.

6. "Assembly Proceedings," September 18, 1787, in ibid., 2:60.

7. About Pennsylvania's representation at the Convention, minority members of the assembly wrote, "We lamented at the time that a majority of our legislature appointed men to represent this state who were all citizens of Philadelphia, none of them calculated to represent the landed [or rural] interests of Pennsylvania, and almost all of them of one political party, men who have been uniformly opposed to [the state] constitution." "The Address of the Seceding Assemblymen," October 2, 1787, in ibid., 2:112.

8. "My Fellow Citizens," *Independent Gazetteer,* November 17, 1787, 3. See also Centinel, "To the Freemen of Pennsylvania," *Independent Gazetteer,* October 5, 1787, 2 (Pennsylvania federalists "flatter themselves that they have lulled all distrust and jealousy of their new plan by gaining the concurrence of the two men in whom America has the highest confidence"); "Northumberland, October 1787," *Pennsylvania Gazette,* October 17, 1787, in DHRC, 2:179 (denying "that a WASHINGTON and his colleagues, whose interests and political salvation are inseparable from ours, would tender a constitution to their brethren fraught with such evils as is by that diabolical [anti-federalist] junto set forth").

9. Centinel, "To the Freemen," 2. The special interests that Centinel referred to in his essay were "the wealthy and the ambitious." Centinel, it is

now known, was Samuel Bryan, who lost his post as assembly clerk when the supporters of the old constitution lost control of that body in 1786.

10. "A Correspondent," *Independent Gazetteer*, October 13, 1787, 2.

11. E.g., "A Federalist," *Independent Gazetteer*, October 10, 1787, 3 ("Not even the *immortal* WASHINGTON, nor the *venerable* FRANKLIN escapes [Centinel's] satire."); "Mr. Printer," *Independent Gazetteer*, October 10, 1787, 3 (condemning Centinel for depicting "*Doctor* Franklin as a *fool* from age, and *General* Washington as a *fool* from nature").

12. One of the People, "To the Freeman of Pennsylvania," *The American Museum* 2 (October 1787): 375. William Livingston was a delegate to the Constitutional Convention from New Jersey.

13. Gouverneur Morris to George Washington, October 30, 1787, PGW-CfS, 5:400.

14. The report first appeared in "Carlisle, January 2," *Carlisle Gazette*, January 2, 1788, 3. It was reprinted eight days later in a newspaper that Washington generally read, the *Pennsylvania Packet*, and subsequently appeared in more than three dozen newspapers, from New Hampshire to Georgia.

15. George Washington to Benjamin Lincoln, January 31, 1788, PGW-CfS, 6:74.

16. E.g., James Madison to George Washington, December 26, 1787, ibid., 5:510.

17. One of the People, "For the Centinel," *Massachusetts Centinel*, October 15, 1787, 34. In a similar manner though now omitting the still uncommitted Hancock, the same source wrote in November about supporting the Constitution, "To mention a WASHINGTON, a FRANKLIN, a MADDISON, a KING and a GORHAM, I think sufficient." One of the People, *Massachusetts Centinel*, "Federal Constitution," November 17, 1787, 69.

18. E.g., Nathan Dane to Henry Knox, December 27, 1787, DHRC, 5:527 ("since I arrived here [Boston] yesterday I find the elections of the province of Main and in the three Western Counties have not been so much in favor of the Constitution as supposed"); Edward Bangs to George Thatcher, January 1, 1788, ibid., 5:571.

19. James Madison to George Washington, February 3, 1788, PGW-CfS, 6:83 (including transcription of a letter from Massachusetts Constitutional Convention delegate Nathanial Gorham).

20. Let Us Think on This, "Federal Constitution," *Massachusetts Centinel*, November 10, 1787, 63.

21. "By Last Night's Mail," *Massachusetts Gazette*, October 19, 1787, 2.

22. E.g., "Extract of a Letter," *Massachusetts Gazette*, January 25, 1788, 3 ("It is free from many of the imperfections with which it is charged," the letter added regarding the Constitution).

23. "The Massachusetts Convention, Monday, 21 January 1788," in DHRC, 6:1287 (Abraham White), 1296 (Amos Singletary), 1297 (Martin Kingsley).

24. Benjamin Lincoln to George Washington, January 27, 1788, PGW-CfS, 6:68.

25. James Madison to George Washington, February 8, 1788, ibid., 6:101 (including transcription of a letter from Massachusetts Constitutional Convention delegate Rufus King).

26. James Madison to George Washington, February 1, 1788, ibid., 6:77 (including transcription of a letter from King).

27. George Washington to John Jay, March 3, 1788, ibid., 6:139.

28. Ibid; George Washington to Henry Knox, March 3, 1788, ibid., 6:140; James Madison to George Washington, February 15, 1788, ibid., 6:115.

29. For a classic portrayal of Hancock, including his dramatic wearing of regal purple during his service as the first president of the Continental Congress following independence, see Herbert S. Allan, *John Hancock: Patriot in Purple* (New York: Macmillan, 1948).

30. "Boston, February 21," *Independent Chronicle* (Boston), February 21, 1788, 3. It was a common practice at such celebrations to make thirteen toasts to correspond with the number of states, but not to make them for those states.

31. "New-York, February 18," *The Daily Advertiser* (New York), February 18, 1788, 2.

32. Rufus King to George Washington, February 6, 1788, PGW-CfS, 6:93–94. Lincoln sent Washington a similar letter on the same day: Benjamin Lincoln to George Washington, February 6, 1788, ibid., 6:94.

33. George Washington to Rufus King, February 29, 1788, ibid., 6:133. Writing to Madison three days later, Washington described the Massachusetts vote as "a severe stoke to the opponents of the proposed Constitution" that "will have a powerful operation on the minds of Men." George Washington to James Madison, March 2, 1788, ibid., 6:136–37.

34. "The Massachusetts Convention, Monday, 4 February 1788," in DHRC, 6:1417 (Thomas Thatcher).

35. The other three states were Georgia (January 2), Connecticut (January 9), and Maryland (April 28). In all four, the support of Washington and Franklin carried influence. E.g., an article in a leading South Carolina

newspaper declared, "God grant that there may be wisdom and goodness enough still found among the *majority* to adopt, without hesitation,
what a WASHINGTON, a FRANKLIN, a MADDISON, &c. so warmly
recommended." "Extract of a Letter," *Columbian Herald,* December 6,
1787, 2. Washington directly intervened in the Maryland ratifying
process by sending letters to key federalists, urging them to proceed
rapidly with ratification as a means to advance prospects elsewhere.
"Postponement of the question would be tantamount to the final rejection of it," Washington advised James McHenry, and "would have
the worst tendency imaginable," he added to Thomas Johnson. In both
letters, Washington noted his past reluctance to "meddle," as he called
it, in "this political dispute," and attributed his change of heart to "the
evils and confusions which will result from the rejection of the proposed Constitution." George Washington to James McHenry, April 27,
1788, PGW-CfS, 6:234; George Washington to Thomas Johnson, April
20, 1788, ibid., 6:218.

36. [Benjamin Franklin] to Editor, *Federal Gazette* (Philadelphia), April 8,
    1788, in PBF, 47:---.

37. See, e.g., George Washington to Henry Knox, March 30, 1788, PGW-
    CfS, 6:183; George Washington to John Langdon, April 2, 1788, ibid.,
    6:187; G. Washington to Jefferson, April 20, 1788, ibid., 6:218.

38. In a letter from this period, Washington observed that in terms of pop-
    ulation and wealth, Virginia "certainly stands first in the Union," but
    added that "Virginians entertain too high an opinion of the importance
    of their" state. George Washington to Bushrod Washington, Novem-
    ber 9, 1787, PGW-CfS, 5:422.

39. George Washington to Benjamin Harrison, September 24, 1787, ibid.,
    5:339. In this letter, Washington wrote that "the political concerns of
    this Country are, in a manner, suspended by a thread" and that if the
    Convention had not agreed on the Constitution, "anarchy would soon
    have ensued." Washington sent the same letter to two other former
    governors, Patrick Henry and Thomas Nelson. Ibid., 5:240.

40. In October, Washington wrote, "It is highly probable that the refusal of
    our Governor and Colo. Mason to subscribe to the proceedings of the
    Convention will have a bad effect in this state." George Washington to
    Henry Knox, October 15, 1788, ibid., 5:376.

41. As early as December 1787, Tobias Lear, Washington's personal secre-
    tary, expressed this concern in a letter from Mount Vernon to Washing-
    ton's old friend, New Hampshire federalist John Langdon. "If I may be

allowed to form an opinion of what would be his wish," Lear wrote of Henry in a message that likely reflected Washington's thinking as well, "it is to divide the Southern States from the others. Should this take place, Virginia would hold the first place among them, & he the first place in Virginia." Tobias Lear to John Langdon, December 3, 1787, DHRC, 8:197. Madison regularly expressed this concern in letters to Washington and others, and it was often raised in Virginia newspapers. E.g., James Madison to Edmund Randolph, January 10, 1787, PJM, 10:355; A Freeholder, *Virginia Independent Chronicle,* April 9, 1788, in DHRC, 9:728.

42. See, e.g., ibid.; George Washington to Benjamin Lincoln, April 2, 1788, PGW-CfS, 6:187; George Washington to John Armstrong, April 25, 1788, ibid., 6:226.

43. "The Virginia Convention, Wednesday, 4 June 1788," in DHRC, 9:929–31 (Patrick Henry).

44. Harlow Giles Unger, *Lion of Liberty: Patrick Henry and the Call to a New Nation* (Cambridge, MA: Da Capo, 2010), 212.

45. "The Virginia Convention, Wednesday, 4 June 1788," in DHRC, 9:931–35 (Edmund Randolph).

46. James Madison to George Washington, June 4, 1788, PGW-CfS, 6:313–14. In April, Washington had reported not knowing the sentiments of the Kentucky members. G. Washington to Lincoln, April 2, 1788, ibid., 6:188.

47. Complaining about "the desultory manner in which [Henry] has treated the subject," at this point, Henry Lee noted that Patrick Henry "seems to have discarded in a great measure, solid argument and strong reasoning, and has established a new system of throwing those bolts, which he has so peculiar a dexterity at discharging." "The Virginia Convention, Monday, 9 June 1788," in DHRC, 9:1072 (Henry Lee). Bushrod Washington made a similar comment in a June 7 letter to George Washington. Bushrod Washington to George Washington, June 7, 1788, PGW-CfS, 6:316 (calling the debates "general and desultory" or random).

48. James Madison to George Washington, June 13, 1788, ibid., 6:326.

49. James Madison to George Washington, June 23, 1788, ibid., 6:351–52.

50. "The Virginia Convention, Tuesday, 24 June 1788," in DHRC, 10:1498 (William Grayson).

51. James Monroe to Thomas Jefferson, July 12, 1788, ibid., 10:1705.

52. George Washington to Charles Cotesworth Pinckney, June 28, 1788, PGW-CfS, 6:361–62; George Washington to Tobias Lear, June 29, 1788, ibid., 6:364.

53. George Washington, "Journal," June 28, 1788, in DGW, 5:351.

54. "PHILADELPHIA, July 4," *Massachusetts Gazette*, July 15, 1788, 2–3.

55. Hugh Blair Grigsby, *The History of the Virginia Federal Convention of 1788* (Richmond: Virginia Historical Society, 1890), 1:157n142. The difficulty in knowing exactly what terms Henry used even in his most quotable speeches is discussed in a 2018 article by John Rogosta, which quotes from Henry's near contemporary biographer, William Wirt: "Even when Henry's speeches were ostensibly recorded contemporaneously by stenographers—for example at Virginia's convention for ratification of the Constitution or during the British Debts Case in federal court—it was reported that stenographers could not follow either 'the captivating flights of Mr. Henry's fancy, or those unexpected and overwhelming assaults which he made on the hearts of his judges.'" John A. Ragosta, "'Caesar Had His Brutus': What Did Patrick Henry Really Say?" *Virginia Magazine of History and Biography* 126 (2018): 182–83 (quoting from Wirt's 1817 biography). Even in 1788, using the "n word" in this setting would qualify as unexpected and overwhelming, but eyewitness accounts confirm it.

56. "The Virginia Convention, Tuesday, 24 June 1788," in DHRC, 10:1476 (Patrick Henry).

57. Ibid.

58. "The Virginia Convention, Tuesday, 17 June 1788," in ibid., 10:1341 (Patrick Henry). With respect to the vulnerability of slavery, George Mason agreed with Henry that Congress "might totally annihilate that kind of property" through taxation. Ibid., 10:1343 (George Mason).

59. "The Virginia Convention, Tuesday, 24 June 1788," in ibid., 10:1503 (James Madison).

60. Ibid., 10:1483 (Edmund Randolph).

61. "The Virginia Convention, Tuesday, 17 June 1788," in ibid., 10:1339 (James Madison).

62. "The Virginia Convention, Tuesday, 24 June 1788," in ibid., 10:1483 (Edmund Randolph).

63. Farrand, May 29, 1787, 1:20–21 (text of Virginia Plan).

64. "The Virginia Convention, Tuesday, 24 June 1788," in DHRC, 10:1504 (Patrick Henry). Earlier that day, Henry had said about Congress and emancipation under the Constitution, "They have the power in clear unequivocal terms." Ibid., 10:1476. Henry was a slippery debater, but first saying the power was express and then (under pressure to prove it) saying it was implied represented a major gaffe. Before this sophisticated audience, Madison beat Henry at every turn.

65. Tench Coxe to James Madison, March 31, 1790, PJM, 13:132.

66. Farrand, August 8, 1787, 2:221–22.

67. Ibid., 2:220.

68. "John Adams' Notes of Debate," July 30, 1776, in *Letters of the Delegates to Congress, 1774–1789,* ed. Paul H. Smith (Washington, D.C.: Library of Congress, 1979), 4:568–69.

69. "The Pennsylvania Convention, Monday, 3 December 1787," in DHRC, 2:463 (James Wilson).

70. Benjamin Franklin, *The Autobiography* (New Haven: Yale University Press, 1964), 69n. (Being treated in a "harsh and tyrannical" manner as an indentured servant, Franklin wrote, "impress[ed] me with that Aversion to arbitrary Power that has stuck to me thro' my whole Life.")

71. Ibid., 43.

72. Benjamin Franklin to John Waring, December 17, 1763, PBF, 10:395. A decade later, Franklin wrote to a French philosopher about freed Blacks in America, "I think they are not deficient in natural Understanding, but they have not the Advantage of Education." Benjamin Franklin to Marquis de Condorcet, March 20, 1774, ibid., 21:151.

73. See Benjamin Franklin, *Observations Concerning the Increase of Mankind, Peopling of Countries, &c* (Boston: Kneeland, 1755), 7 ("The Whites who have slaves, not labouring, are enfeebled," Franklin wrote, "the slaves being work'd too hard, and ill fed, there constitutions are broken").

74. Among the early Philadelphia-area Quaker abolitionists whom Franklin published were four of the most influential: Ralph Sandiford, Benjamin Lay, John Woolman, and Anthony Benezet. In 1772, speaking about their shared opposition to the slave trade, Benezet called Franklin "a real friend and fellow traveller on a dangerous and heavy road." Anthony Benezet to Benjamin Franklin, April 27, 1772, PBF, 19:113. Franklin's experience as a young printer publishing ads for slaves is somewhat similar to Woolman's experience as a young shopkeeper participating in the sale of slaves, about which Woolman wrote, "My Employer having a Negro Woman, sold her, and desired me to write a Bill of Sale, the Man being waiting who bought her: The Thing was sudden; and, though the Thoughts of writing an Instrument of Slavery for one of my Fellow-creatures felt uneasy, yet I remembered I was hired by the Year, that it was my Master who directed me to do it, and that it was an elderly Man, a Member of our Society, who bought her; so, through Weakness, I gave way." John Woolman, *The Journal and Other Writings* (London: Dent, 1910), 27. As these comments sug-

gest, the institution of slavery was so widely accepted among whites in Pennsylvania, including members of the Society of Friends (or Quakers), that Woolman and Franklin acquiesced.

75. Benjamin Franklin, "Apology for Printers," *Pennsylvania Gazette,* June 10, 1731, in PBF, 1:195.

76. Benjamin Franklin to Deborah Franklin, June 10, 1758, ibid., 8:92.

77. In a representative assertion from the period, quoted in part in chapter 5, Franklin wrote to a longtime confidant in England about American independence, "I am too well acquainted with all the Springs and Levers of our Machine, not to see that our human means were unequal to our undertaking, and that if it had not been for the Justice of our Cause, and the consequent Interposition of Providence in which we had Faith we must have been ruined. If I had ever before been an Atheist I should now have been convinced of the Being and Government of a Deity." Benjamin Franklin to William Strahan, August 19, 1784, ibid., 42:29.

78. Benjamin Franklin to John Lathrop, May 31, 1788, ibid., 47:---.

79. Richard Henry Lee to George Washington, April 6, 1789, PGW-PS, 2:29.

80. "Philadelphia, 20 April," *Federal Gazette,* April 20, 1789, 2; "Philadelphia, April 29," *Freemans's Journal,* April 29, 1789, 3; "Philadelphia, 22 April," *New-York Daily Gazette,* April 27, 1789, 410; William Spohn Baker, *Washington after the Revolution* (Philadelphia: Lippincott, 1898), 124.

81. Annals of Congress (Senate), April 30, 1789, 1:28.

82. Virginia governor Thomas Jefferson estimated that thirty thousand slaves fled behind British lines during the 1781 invasion of Virginia. See Henry Wiencek, *An Imperfect God: George Washington, His Slaves, and the Creation of America* (New York: Farrar, Straus and Giroux, 2003), 203.

83. Benjamin Franklin, *Observations Concerning the Increase of Mankind,* 4th ed. (1769), in PBF, 19:133n1. In the original 1751 privately circulated manuscript (first published in 1754), this passage read, "Almost every Slave *by Nature* a thief." Benjamin Franklin, "Observations Concerning the Increase of Mankind" (1751), in PBF 4:229. Clearly Franklin had changed his view of slaves. Regarding Washington, in his Pulitzer Prize–winning biography, Ron Chernow called him "a tough master," wrote of his punishing "difficult" slaves by selling them to their doom in the West Indies, and noted that "on his plantation, Washington demanded high performance and had little patience with sluggards and loafers." Ron Chernow, *Washington: A Life* (New York: Penguin Press, 2010), 640.

84. George Washington to Tobias Lear, April 12, 1791, PGW-PS, 8:85.

85. George S. Brookes, *Friend Anthony Benezet* (Philadelphia: University of Pennsylvania Press, 1937), 162.

86. Benjamin Franklin to John Langdon, October 25, 1788, PBF, 47:---. In his response to Franklin's letter, Randolph wrote "merely as a private man," and not in an official capacity, "that whensoever an opportunity shall present itself, which shall warrant me, as a *citizen,* to emancipate the slaves, possessed of me," he would do so "regardless of the loss of property." Edmund Randolph to Benjamin Franklin, August 2, 1788, ibid., 47:---. Randolph argued against the slave trade at the Constitutional Convention but never freed his own slaves.

87. Josiah Wedgwood to Benjamin Franklin, February 29, 1788, ibid., 47:---.

88. Benjamin Franklin to Marquis de Lafayette, July 7, 1788, ibid., 47:---.

89. Granville Sharp to Pennsylvania Abolition Society, July 30, 1788, ibid., 47:---.

90. Benjamin Franklin to the Senate and House of Representatives of the United States, February 3, 1790, ibid., 48:---.

91. One critical congressman noted that the petitions were signed only by persons calling themselves a clerk. *Lloyd's Notes,* February 12, 1790, in DHFFC, 12:297 (William Smith). Another warned, "There never was a [religious] society of any considerable extent which did not interfere with the concerns of other people, and this interference has at one time or other deluged the world in blood." *Gazette of the United States,* February 17, 1790, in ibid., 12:305 (Michael Stone).

92. Annals of Congress (House), February 12, 1790, 1:1240 (Tucker), 1:1242 (Baldwin); ibid., February 11, 1790, 1:1231 (Gerry), 1:1229 (Jackson). Addressing opponents' argument about religious establishment, Thomas Scott of Pennsylvania stated, "I look upon the slave-trade to be one of the most abominable things on earth; and if there was neither God nor devil, I should oppose it upon the principles of humanity." Ibid., February 12, 1790, 1:1241. Many critical members spoke of the economic imperative of slaves. E.g., *Lloyd's Notes,* February 12, 1790, in DHFFC, 12:296 (James Jackson: "rice cannot be brought to market without these people"); *Gazette of the United States,* February 17, 1790, in ibid., 12:303–4 (William Smith: "if there were no slaves in the southern States they would be entirely depopulated—for from the nature of the country it could not be cultivated without them").

93. *New-York Daily Gazette,* February 15, 1790, in DHFFC, 12:302.

94. *Gazette of the United States,* February 17, 1790, in ibid.

95. Ibid., 12:304. As an example of how Congress could constitutionally un-

dermine slavery despite the bar on banning the slave trade until 1808, Madison observed that it might make some regulations "in relation to the introduction of [slaves] into the new States to be formed out of the Western Territory." Annals of Congress (House), February 12, 1790, 1:1246.

96. Benjamin Franklin to George Washington, September 16, 1789, PGW-PS, 4:47–48.

97. George Washington to Benjamin Franklin, September 23, 1789, ibid., 4:66.

98. Benjamin Franklin to Louis-Guillaume Le Veillard, September 5, 1789, PBF, 46:---.

99. Benjamin Franklin to Benjamin Vaughan, November 2, 1789, in Jared Sparks, ed., *The Works of Benjamin Franklin* (Philadelphia: Childs & Peterson, 1840), 10:397.

100. For "much emaciated," see Thomas Jefferson to Ferdinand Grand, April 4, 1790, PTJ, 16:298; for "almost too much," see Thomas Jefferson, "Autobiography," January 6, 1821, in PTJ-RS, --:---. Reflecting his mental sharpness, in response to a question by Jefferson (as secretary of state) about an obscure point in the peace treaty with Britain, Franklin (as the main architect of that treaty) sent Jefferson a detailed answer on April 8, closing with the words, "I have the Honor to be with the greatest Esteem & Respect Sir, Your most obedient & most hble Servant." It was Franklin's last letter. Benjamin Franklin to Thomas Jefferson, April 8, 1790, PTJ, 16:326.

101. *New-York Daily Gazette,* March 17, 1790, in DHFFC, 12:723.

102. *Lloyd's Notes,* March 15, 1790, in ibid., 12:719; *Daily Advertiser,* March 18, 1790, in ibid., 12:732. About race mixing, Jackson declared, "However fond the Quakers may be of this mixture and of giving their daughters to negro sons, and receiving the negro daughters for their sons, there will be those who do not approve of the breed, and a motley breed it will be." *Daily Advertiser,* March 18, 1790, in ibid., 12:728.

103. "Report of the Special Committee" and "Report of the Committee of the Whole House," Annals of Congress (House), March 23, 1790, 1:1523–25.

104. Annals of Congress (House), March 23, 1790, 1:1523.

105. "The final decision" on the petitions, Washington wrote in a letter to his in-law and confidant David Stuart, "was as favourable as the proprietors of that species of property could well have expected considering the great dereliction to Slavery in a large part of this Union." George Washington to David Stuart, June 15, 1790, PGW-PS, 5:525. Washington, of course, was a proprietor of that species of property.

106. David Stuart to George Washington, March 15, 1790, ibid., 5:236.

107. George Washington to David Stuart, March 28, 1790, ibid., 5:286–88 (the bracketed word is unintelligible in the original—"likely" fits in context, but it could be a stronger term, such as "possible").

108. Historicus [Benjamin Franklin] to the Editor of *Federal Gazette,* March 23, 1790, PBF, 48:---.

109. Ibid.

110. Ibid.

111. G. Washington to B. Franklin, September 23, 1789, 4:66.

112. Benjamin Franklin to William Vaughan, December 9, 1788, PBF, 47:---.

113. Benjamin Franklin to Jane Mecom, October 19, 1789, ibid., 48:---.

114. Thomas Jefferson to Benjamin Rush, October 4, 1803, PTJ, 41:471. In particular, Jefferson singled out John Adams ("whom he certainly did not love") as one person whom Washington would not want so honored.

115. G. Washington to Stuart, June 15, 1790, 5:525.

116. National Assembly of France to George Washington, June 20, 1790, PGW-PS, 5:541.

117. George Washington to President of the National Assembly of France, January 27, 1791, ibid., 7:292.

Epilogue: The Walking Stick

1. Henry Hill to George Washington, May 7, 1790, PGW-PS, 5:388 (quoting from the codicil to Franklin's will). In the original French, which Franklin employed in writing the 1789 codicil, the actual phrase was *"le baton de Pommier savage,"* or "wild apple tree," rather than the "fine Crab tree" translation provided by his executor in the transmittal letter to Washington. But *"mon ami"* and *"un sceptre"* were there as translated from Franklin's original codicil. Benjamin Franklin, Codicil to Will, June 23, 1789, in PBF, 47:---.

2. Hill to G. Washington, May 7, 1790, 5:389n1 (quoting from cover letter to Franklin accompanying the original gift); Franklin, Codicil, June 23, 1789, 47:---.

3. George Washington to Henry Hill, June 3, 1790, PGW-PS, 5:389n1.

4. For Adams's proposed titles, see "Diary of William Maclay," April 24 and May 8, 1789, in DHFFC, 9:4–5, 28–29; James H. Hutson, "John Adams' Title Campaign," *New England Quarterly* 41 (1968): 31–34. These proposed titles appear in private letters but at least one senator noted that Adams also suggested titles in a speech to the Senate. "Diary of Maclay," May 8, 1789, 9:28.

5. George Washington, "To the PEOPLE of the United States," *Claypoole's American Daily Advertiser*, September 19, 1796, 2.

6. Tobias Lear, "Narrative Accounts of the Death of George Washington: Diary Account," December 14–15, 1799, in PGW-RS, 4:542.

7. Ibid., 4:545. Fearing that they might kill her to gain their freedom, Martha Washington freed her husband's slaves one year after his death.

8. "Boston, December 28, 1799," *Columbian Centinel*, December 28, 1799, 2.

9. Robert Pleasants to George Washington, December 11, 1784, PGW-CfS, 3:449.

10. Jennifer Schuessler, "History and Memory Are Not Set in Stone," *New York Times*, August 15, 2017, A12. In 2019, the San Francisco Board of Education voted to paint over a federally funded New Deal Era Works Progress Administration mural painted by radical artist Victor Arnautoff that critically depicted Washington's westward expansion policy. The mural was painted in 1936 for the board's George Washington High School and includes the image of Franklin in a group of four founders, with Washington dispatching settlers westward over the corpse of a Native American. Franklin appears skeptical about the operation. Derrick Bryson Taylor, "School Will Cover Up W.P.A.-Era Murals," *New York Times*, June 29, 2019, A15.

11. Pleasants to G. Washington, December 11, 1784, 3:450.

12. [Benjamin Franklin], *Poor Richard, 1735*, in PBF, 2:9; [Benjamin Franklin], *Poor Richard, 1736*, in ibid., 2:142.

13. See Paul K. Longmore, *The Invention of George Washington* (Berkeley: University of California Press, 1988), 181.

14. John Adams to Benjamin Rush, March 19, 1812, in *The Spur of Fame: Dialogues of John Adams and Benjamin Rush, 1805–1813*, eds. John A. Schultz and Douglas Adair (San Marino, Calif.: Huntington, 1966), 211–12.

15. In this role Washington played the part of a Virginia gentleman and, as the proud New Englander John Adams added when writing about it, "Virginian geese are all swans." John Adams to Benjamin Rush, November 11, 1807, in *Old Family Letters*, ed. Alexander Biddle (Philadelphia: Lippincott, 1892), 169.

16. Farrand, June 2, 1787, 1:83.

17. Edward R. Murrow, "A Report on Senator Joseph R. McCarthy," *See It Now*, CBS, March 9, 1954.

18. Franklin D. Roosevelt, 1941 State of the Union Address, 87 Cong. Rec.—House, pt. 1, p. 45 (January 6, 1941).

INDEX